An Introduction to
Mathematical Billiards

An Introduction to **Mathematical Billiards**

Utkir A. Rozikov

V. I. Romanovskiy Institute of Mathematics, Tashkent, Uzbekistan

NEW JERSEY · LONDON · SINGAPORE · BEIJING · SHANGHAI · HONG KONG · TAIPEI · CHENNAI · TOKYO

Published by

World Scientific Publishing Co. Pte. Ltd.

5 Toh Tuck Link, Singapore 596224

USA office: 27 Warren Street, Suite 401-402, Hackensack, NJ 07601

UK office: 57 Shelton Street, Covent Garden, London WC2H 9HE

Library of Congress Cataloging-in-Publication Data

Names: Rozikov, Utkir A., 1970–   author.

Title: An introduction to mathematical billiards / by Utkir A. Rozikov.

Description: New Jersey : World Scientific, [2018] | Includes bibliographical references and index.

Identifiers: LCCN 2018040369 | ISBN 9789813276468 (hardcover : alk. paper)

Subjects: LCSH: Differentiable dynamical systems. | Billiards.

Classification: LCC QA614.8 .R68 2018 | DDC 515/.39--dc23

LC record available at https://lccn.loc.gov/2018040369

British Library Cataloguing-in-Publication Data

A catalogue record for this book is available from the British Library.

First published 2019 (Hardcover)
Reprinted 2020 (in paperback edition)
ISBN 978-981-122-125-5 (pbk)

For any available supplementary material, please visit
https://www.worldscientific.com/worldscibooks/10.1142/11162#t=suppl

To the memory of my father

Abdullo Rozikov (1944-2017)

My father was an excellent teacher of physics and he had an
exceptional talent in physics.

Contents

Preface

A *mathematical billiard* is a mechanical system consisting of a billiard ball on a table of any form but without billiard pockets. On this table the ball moves without friction, and it is absolutely elastic when reflected from boards (the boundary of the table). The *mathematical problem* of billiards is to investigate the behavior of a trajectory of this ball.

The purpose of this book is to systematically present known results on the behavior of a billiard ball on a planar table (having one of the following forms: circle, ellipse, triangle, rectangle, polygon and some general convex domains).

The description of these trajectories leads to the solution of various questions in mathematics and mechanics: problems about liquid transfusion, lighting of mirror rooms, crushing of stones in a kidney, collisions of gas particles etc.

The analysis of billiard trajectories can involve methods of geometry, dynamical system, and ergodic theory, as well as methods of theoretical physics and mechanics.

In July of 2018, MathSciNet found more than 2420 entries for "billiards" in the entire database. By this database the first publication related to billiards is [75], which was published in 1905. In just 2017 alone, there were 63 publications. Therefore billiards is a very popular topic in mathematics.

Billiards is mainly studied in the framework of the theory of dynamical systems. The number of young scientists interested in mathematical billiards is increasing because billiards has many nice applications in biology, mathematics, medicine, and physics. The long list of literature devoted to billiards makes it difficult for a beginner to start reading the theory. The main aim of this book is to help the reader by giving a systematic review of the theory of dynamical systems, presentation of billiards in elementary mathematics and simple billiards related to geometry and physics. There are a number of surveys devoted to mathematical billiards, from popular to technically involved: [37], [38], [48], [66], [72], [108], [109], [110], [111] etc. But in this book I used some Russian materials (not available in English), modern papers and internet sources which were not previously found in textbooks on billiard theory. I am sure that the book will be very helpful for readers interested in dynamical systems and mathematical billiards.

My interest in mathematical billiards started when, as a student of Samarkand University, I was reading [37], which is a Russian book published in 1990. After reading the book [37] I was able to convince people (who were not mathematicians and who asked me about the usefulness of mathematics in real life) that mathematics is very beautiful, rich, and useful.

I tried to popularize the theory of mathematical billiards when I was teaching in Bukhara, Karshi, and Namangan universities of Uzbekistan. Moreover, I had many discussions on the billiard theory when I was abroad (mainly with non-Russian speakers). After such discussions I planned to write this book to serve as a basic introduction to the theory of mathematical billiards.

The book is based on materials collected during my several visits to the University of Cambridge, Newton Institute and University of Leeds, UK; International Center for Theoretical Physics, Trieste, Italy; "La Sapienza" University, Rome, Italy; IHES, Bures-sur-Yvette, France; Universite du Sud Toulon Var, Centre de Physique Theorique and IMéRA, Marseille, France; University Paris-Est, France; University Santiago de Compostela and University Granada, Spain; University of Bonn and Ruhr University, Bochum, Germany.

Acknowledgements. I thank all the above-mentioned institutions for their warm hospitality and excellent working conditions. I am very grateful to the American students A.J.M. Hardin, A.R. Luna and S.Yam for checking the grammer and language of this book. I thank World Scientific Publishing for the opportunity to publish this book.

I am indebted to my family for their warm attitude to my work. My sons (Azamat and Laziz) have helped me to draw the many figures in this book.

Utkir A. Rozikov
Tashkent, Uzbekistan
August 2018

Introduction

The mathematical study of billiards generally consists of one billiard ball on a table Q of arbitrary form (which can be planar or even a multidimensional domain). The ball moves and its trajectory is defined by the ball's initial position, $q \in Q$, and its initial speed vector, $v = v(0)$. The ball's reflections from the boundary of the table are assumed to satisfy this law: the reflection and incidence angles are the same. The absolute value of the speed $v(t)$ of the ball at time t is assumed to be constant. In the case when the table has a curved boundary then the mentioned angles will be considered with respect to the tangent line at the point (if the tangent exists). If at this point the tangent line does not exist (non-unique) then the ball is assumed to stop there, or alternatively it can be assumed that the ball goes back by the same trajectory. The main aim is to investigate the trajectory of a ball on such a table - not only for a few bounces, but over a very long time.

In the detailed study of billiards many interesting results can arise. For example, any smooth planar convex set has at least two 2-periodic trajectories. Additionally there are always $\phi(n)$ distinct n-gonal periodic orbits (where $\phi(n)$ is Euler's function, i.e., it counts the positive integers up to a given integer n which are relatively prime to n) on a smooth billiard table. Over the past 40 years, billiard theory has been connected with number theory, geometry, spectral theory, acoustics, optics, thermodynamics and many other branches of physics. Therefore mathematical billiards represents one of the most popular classes of dynamical systems. As a dynamical system, mathematical billiards demonstrates a wide variety of behaviors, including regular, periodic, chaotic and mixed phase space dynamics.

In Chapter 1 we briefly give the theory of discrete and continuous-time dynamical systems, this will be helpful for the reader, because a mathematical billiard which we want to discuss in this book is a particular case of dynamical systems. In the last section of this chapter we define a mathematical billiard and formulate main problems related to them.

Today there are more than 2400 published works devoted to the theory of mathematical billiards. Here we give a very short review of billiards considered on planar tables. Since there are more than 2000 papers devoted to such planar billiards, my review is not a complete one. For more survey and historical notes on mathematical billiards see [66], [49]. I will classify the review by the classification of billiard tables:

1. Circular table. Since the circle has rotational symmetry, a billiard trajectory is completely determined by the angle θ made with the circle. This angle remains the same after each reflection. If θ is a rational number, then every trajectory of the billiard ball is periodic. Moreover, if $\theta = p/q$ then the length of periodicity is equal to q. If θ is an irrational number, then every trajectory is dense on the circle. It is known that any billiard trajectory in a circle never comes within some concentric circle in the boundary of which all segments of the trajectory are tangent lines [37], [38], [86]. These results are presented in Chapter 2.

2. Elliptical table. Billiards on an elliptical table is completely understood ([37], [86], [110]): In general, a billiard trajectory inside an ellipse remains tangent to a confocal conic. Moreover, depending on the initial position of the ball we have the following cases:

2.1. If the billiard ball passes over a focus of the ellipse then the ball hits the border of the table and will bounce back along the second focus. After a new hit the ball will bounce through the other first focus etc. In the future the trajectory will be very close to the major axis.

2.2. If the billiard trajectory crosses the line segment between the two focuses of the ellipse, then after every bounce the trajectory will intersect with the line segment between the focuses. Then the line segment of the trajectory is tangent to the hyperbola which shares the same foci with the ellipse.

2.3. If the trajectory of billiard initially does not intersect with the line segment between the two focuses of the ellipse, then the trajectory of the billiard is tangent to an ellipse which shares the same foci with the given table. In Chapter 3 we give proofs of these results.

3. Triangular table. Billiard tables with a non-smooth boundary can be a planar polygon. In particular, some basic physical models concern the billiard on triangles. Such billiards were investigated for example in [26],

[37], [39] [40], [47], [51], [52], [44], [57]-[59], [68], [98], [99], [117], [118], [119] and more recently in [96], [113], [114].

In these papers, in particular, the following results were obtained. In every acute triangle there exists a periodic trajectory with three links (which is called the Fagnano trajectory) and a bundle of 6-link trajectories parallel to it. The set of the acute triangles were divided into infinitely many regions, each triangle of which contain periodic trajectories with some odd (≥ 3) number of links. A set of obtuse triangles is found with the property that every triangle of the set contains a periodic trajectory.

In [39] some (mirror) periodic trajectories in irrational right triangles were given. It is also known that all rational triangles (i.e. all their angles are rational multiples of π) contain periodic trajectories. Moreover, any perpendicular trajectory (it meets a side of the triangle being perpendicular to it) in a rational triangle is periodic except if it hits a vertex of the polygon (singularity). In [14], [15] it is proven for every rational triangle that periodic points of the billiard are dense in the phase space of the billiard. The last fact is a strengthening of results in [77]. In [26] it is proved that almost all perpendicular trajectories in a right triangle are periodic. In [41] it is shown that through every point of every right triangle there passes a periodic billiard trajectory. Moreover, these periodic trajectories are not stable, meaning that they can be destroyed by a small perturbation of the triangle.

The only periodic billiard trajectory of a single circuit in an acute triangle is the pedal triangle. There are an infinite number of multiple-circuit paths, but all segments are parallel to the sides of the pedal triangle. In general the following problem is unsolved: Does every obtuse triangle have a periodic orbit? Another question is do all tables admit a periodic orbit? These are difficult problems. The best known result is that every triangle with no angle more than 100 degrees will have a periodic billiard trajectory [98], [99]. Despite the numerous publications devoted to the billiards on triangles there are still many open problems. Nowadays, this direction is actively developing. (Chapter 3 contains some of the above mentioned results in detail.)

4. Polygonal table. One can also consider billiard trajectories on polygonal billiard tables, see for example [13], [14], [15], [31], [37], [39], [40], [45], [51], [70], [106], [108]-[111], [115], [118], [121], [125] and the references therein. Everything above mentioned for rational triangles is also true for rational polygons. In the study of polygonal billiards (in particular, triangular billiards) one powerful and simple method is related to the unfolding of the trajectory: when a trajectory strikes a boundary, the method of unfolding does not have the trajectory reflected back into the billiard table, instead one reflects the billiard table across the boundary where the ray struck. The trajectory then crosses the boundary into the reflected image of the billiard table. The same process happens within the reflected image of the billiard table when the ray strikes another boundary. Thus unfolding any trajectory becomes expressed as a straight line. This method is used to show which polygons tile the plane with reflections, moreover, periodic trajectories in polygonal billiards are studied with this method. There some open questions in the field of polygonal billiards too. For example, do all polygons have a periodic orbit? Is it possible to give all kind of periodic trajectories for a polygonal table? These billiards are also discussed in Chapter 3.

5. General convex table. G.D. Birkhoff was first to consider billiards systematically as models for problems of classical mechanics. Birkhoff considered billiards only in smooth convex domains see, for example, [31], [37], [38], [39], [49], [74], [110], [111].

Recall here a result of J. Birkhoff [39]: For a billiard on an arbitrary planar convex domain (table), with a bounded, closed and smooth boundary, there exists a periodic trajectory of the length n (for any given $n \geq 2$).

In [7] Rogers-Shephard type inequalities are proved for the shortest closed billiard trajectories in general convex domains. For hexagon-triangle and Hanner polytopes it is shown that any regular closed billiard trajectory is of minimal length and such orbits are abundant. Moreover, for a planar domain of constant width it is shown that any minimal closed billiard trajectory is 2-periodic. For more details about billiards in general convex domains see Chapter 3 and references mentioned above.

Chapter 4 of this book deals with how mathematical billiards can arise in many problems of physics. For example, one may consider billiards in potential fields. Another interesting modification, popular in the physical literature, is the billiard in a magnetic field [11], [110], [111]. There we give some simple examples from physics which are related to a billiard: Motion and collisions of balls, Mechanical interpretations of three-periodic points, Billiard trajectories of light, Corner reflector, Crushing of stones in kidneys etc. This book does not discuss applications of the billiard to the billiard-ball computer which played a significant role in the development of the quantum computer and its application to semiconductor device physics.

As noted above this book presents an introduction to the theory of mathematical billiards. There is a huge part of the theory which is not discussed here because the theory of mathematical billiards can be partitioned into three areas [78]: convex billiards with smooth boundaries, billiards in polygons and polyhedra, and dispersing and semi-dispersing billiards. These areas differ by the types of results and the methods of study: in the former a prominent role is played by the KAM theory and the theory of area preserving twist maps; the latter concerns hyperbolic dynamics and has much in common with the study of the geodesic flow on negatively curved manifolds. The recent progress in the study of polygonal billiards is mostly due to applications of the theory of flat structures on surfaces and the study of the action of a Lie group and generalizations (see [24], [25], [46], [51], [64] [65], [66], [71], [72], [104], [109]-[111], [122], [118], [124]). In [3] a notion of symplectic billiards is introduced and its basic properties are studied. The book [37] is written in an accessible manner, and touches upon a broad variety of questions. This book can undoubtedly provide pleasurable and instructive reading for any mathematician or physicist interested in billiards, dynamical systems, ergodic theory, mechanics, geometry, partial differential equations, and mathematical foundations of statistical mechanics.

An introduction to the theory of billiards for a more advanced reader can be found in [27], [72] and the next level is represented in the book [108]. The book [65] contains an presentation of the theory of convex billiards and twist maps. The theory of parabolic billiards is contained in a survey paper [78]. The volume [107] contains rich material on hyperbolic billiards and related questions.

For many open problems of the theory of mathematical billiards see [50], [49]. In this book many exercises are given.

Chapter 1

Dynamical systems and mathematical billiards

In this chapter we briefly give the theory (without proofs) of discrete and continuous-time dynamical systems. This will be helpful for the reader, because a mathematical billiard which we want to discuss in this book is a particular case of dynamical systems. We define discrete and continuous-time dynamical systems, formulate the main problem in consideration of such a dynamical system and give a review of methods of solutions of the main problem. In the last section of this chapter we define mathematical billiards and formulate main problems related to them.

1.1 Discrete-time dynamical systems

1.1.1 *Definitions and the main problem*

What do we mean by a system? A *system* is a set of interacting or interdependent components, members or points.

For example, let S be the members of a family, i.e., $S = \{parents, childeren\}$, this set is a system, because there are some relations (interactions) between them. Consider now a set which is $\{chalk, coin, pen, key\}$ this is a set but is not a system, because there is no interaction between these objects.

The state of a system is a collection of its properties, that are interesting to know.

For our example, the state of S (the family) can be considered as $s =$ the number of boys, the number of schoolboys, salary of the father etc.

Measure of a system is the assignment of numbers to a state or a property of the system. It is a cornerstone of most natural sciences, technology, economics, and quantitative research in other social sciences.

1

For example, we can give the following distinct measures of the system S.

$M_{sport}(S) =$ the number of boys in S.

$M_{school}(S) =$ the number of schoolboys in S.

Let us assume that a thief robs the family, then the following measure is interesting to the thief:

$M_{thief}(S) =$ Total sum of money in S.

So one can define different measures of the same system S.

A dynamical system is a system the state of which changes when time is increasing.

Again the system S (the above mentioned family) is a dynamical system because, states of S change when time increases: for example, the number of schoolboys, the salary of the father etc.

We note that, in case when a system contains infinitely many elements, then, investigation and measurement of such dynamical systems are very difficult and require powerful mathematical methods. You can just imagine how the system S will be complicated if there are infinitely many people in it. What will be $M_{thief}(S)$ in the case where S is infinite? Is it still a finite number? If it is infinite, then the thief will die from a heart attack!

Thus a dynamical system is a rule that describes the evolution with time of a point in a given set. This rule might be specified by very different means like iterated maps (discrete-time), ordinary differential equations and partial differential equations (continuous-time) or cellular automata. There are some dynamical systems where instead of time parameters, one can consider the changes of a system depending on other parameters, like temperature (making thermodynamic systems) etc.

The main problem of mathematicians for a given dynamical system: is to know all changes of states (evolution) of the system when time goes to infinity. This evolution can occur smoothly over time or in discrete time steps.

In this section, we consider dynamical systems where the state of the system evolves in discrete time steps (i.e., discrete dynamical systems).

In order to complete the description of the dynamical system, one needs to fix a rule that determines, given an initial state, what the resulting sequence of future states must be.

Note that if the time t is continuous, then the dynamical system is defined by the rule of a differential equation whose independent variable is time, then the main problem is to know the behavior of solutions of the equation in the distant future ($t \to \infty$) or the past ($t \to -\infty$).

Using MathSciNet search, one can see that there are about more than 830 *books* devoted to theory of dynamical systems. The rest of this section is mainly based on books [33], [35], and [36].

To define a discrete-time dynamical system we consider a function $f : X \to X$. At the moment we do not give conditions of f, the only assumption is that f is from X to itself.

For $x \in X$ denote by $f^n(x)$ the n-fold composition of f with itself (i.e. n time iteration of f to x):

$$f^n(x) = \underbrace{f(f(f \ldots (f(x))) \ldots)}_{n \text{ times}}.$$

Definition 1.1. For arbitrary given $x_0 \in X$ and $f : X \to X$ the discrete-time dynamical system (also called forward orbit or trajectory of x_0) is the sequence of points

$$x_0, x_1 = f(x_0), x_2 = f^2(x_0), x_3 = f^3(x_0), \ldots \tag{1.1}$$

Definition 1.2. A point $x \in X$ is called a fixed point for $f : X \to X$ if $f(x) = x$. The point x is a periodic point of period p if $f^p(x) = x$. The least positive p for which $f^p(x) = x$ is called the prime period of x.

Denote the set of all fixed points by $\mathrm{Fix}(f)$ and the set of all periodic points of (not necessarily prime) period p by $\mathrm{Per}_p(f)$.

It is clear that the set of all iterates of a periodic point form a periodic sequence (orbit).

Example 1.1. For $f(x) = -x$ we have $\mathrm{Fix}(-x) = \{0\}$, $\mathrm{Per}_2(-x) \setminus \mathrm{Fix}(-x) = \mathbb{R} \setminus \{0\}$.

For $f(x) = x^2$ we have $\mathrm{Fix}(x^2) = \{0, 1\}$, $\mathrm{Per}_2(x^2) \setminus \mathrm{Fix}(x^2) = \emptyset$.

Denote the sequence (1.1), for $x = x_0$, by $\tau(x) = \{x_j = f^j(x), j = 0, 1, \ldots \}$.

Exercise 1. Find fixed and periodic points for the following functions:
   a) $f(x) = |x|$,   b) $f(x) = x^3$,   c) $f(x) = -x - 1$,
   d) $f(x) = \sin(x)$,   e) $f(x) = \tan(x)$,   f) $f(x) = x(x^2 - 3x + 2) - 1$.

Definition 1.3. The limit set $w(x_0)$ of orbit (1.1) is defined as

$$w(x_0) = \cap_{j=1}^{\infty} \overline{[\tau(f^j(x_0)]},$$

where $\overline{M}$ is the closure[1] of M.

---

[1] The closure of a set M of points (in a topological space) consists of all points in M together with all limit points of M. Alternatively, the closure of M is the union of M and its boundary.

Definition 1.4.

- A fixed point x^* of f is attracting or stable if there is an open set (neighborhood) U containing x^* such that for all $x \in U$, the orbit $\tau(x)$ with initial point x converges to x^*.
- A fixed point x^* of f is repelling if there is an open set (neighborhood) U containing x^* such that for any $x \in U$, $x \neq x^*$, there is some $k \geq 1$ such that $f^k(x) \notin U$.

The main problem: Given a function f and initial point x_0 what ultimately happens with the sequence (1.1). Does the limit $\lim_{n \to \infty} x_n$ exist? If not what is the set of limit points of the sequence? Is this set finite or infinite?

The difficulty of the main problem depends on the set X and on the given function f. The problem is mainly considered in case $X \subset \mathbb{R}^m$, $m \geq 1$ and when f is a continuous function on X. Let us give a list of known results.

1.1.2   *One-dimensional systems*

Consider a dynamical system where X is a subset of $\mathbb{R}$ and f is a continuous function on X. Assume that f is also continuously differentiable on X.

Definition 1.5. A fixed point $x^* \in X$ is called hyperbolic if $|f'(x^*)| \neq 1$.

The following theorem is well-known and gives the main tool to check the type of the fixed point.

Theorem 1.1. *Let $X \subset \mathbb{R}$ and f be continuously differentiable on X. Let $x \in X$ be a hyperbolic fixed point of f then*

1) If $|f'(x^*)| < 1$, then x^* is attracting.
2) If $|f'(x^*)| > 1$, then x^* is repelling.

Example 1.2. Consider $X = \mathbb{R}$ and $f(x) = ax + b$. For this function we have

$$\mathrm{Fix}(f) = \begin{cases} \{\frac{b}{1-a}\}, & \text{if } a \neq 1 \\ \mathbb{R}, & \text{if } a = 1, b = 0. \end{cases}$$

Moreover, the case $a = 1, b = 0$ is not interesting, because in this case $f(x) = id$. Therefore, we consider $a \neq 1$. Since $f'(x) = a$, by Theorem

1.1 we have that the fixed point $\frac{b}{1-a}$ is attracting if $|a| < 1$ and repelling if $|a| > 1$.

The linearity of the function $f(x) = ax + b$ allows to simply compute all iterations f^n:

$$x_n = f^n(x_0) = a^n x_0 + b \sum_{i=0}^{n-1} a^i = \begin{cases} a^n x_0 + \frac{b(1-a^n)}{1-a}, & \text{if } a \neq 1 \\ x_0 + bn, & \text{if } a = 1. \end{cases} \tag{1.2}$$

To find p-periodic points of f we solve $f^p(x) = x$, where $p \geq 2$ is a natural number. Here we consider the non-trivial case $a \neq 1$, then by (1.2) the equation $f^p(x) = x$ has the following form

$$a^p x + \frac{b(1 - a^p)}{1 - a} = x.$$

From this equation for $a \neq -1$ (recall $a \neq 1$) we get $x = \frac{b}{1-a}$, i.e., if $a \neq \pm 1$ then only the unique fixed point is p-periodic. In case $a = -1$ the equation is reduced to

$$((-1)^p - 1)\left(x - \frac{b}{2}\right) = 0, \tag{1.3}$$

which has unique solution $x = \frac{b}{2}$ (the fixed point) if p is an odd number. But in case when p is an even number then any $x \in \mathbb{R}$ is a solution to (1.3).

For $a = -1$ we have $f^2(x) = x$ has two solutions $x = \alpha$ and $x = b - \alpha$, for any $\alpha \in \mathbb{R}$.

Thus independently on the value of the *even* number $p = 2, 4, 6, \ldots$ we have

$$\mathrm{Per}_p(f) \setminus \mathrm{Fix}(f) = \begin{cases} \emptyset, & \text{if } a \neq -1 \\ \{x, b - x\}_{x \in \mathbb{R}}, & \text{if } a = -1. \end{cases}$$

Note that the prime period of each element of $\{x, b-x\}_{x \in \mathbb{R} \setminus \{b/2\}}$ for $a = -1$ is $p = 2$.

Using formula (1.2) we get the following limit (full answer to the main problem of the dynamical system):

$$\lim_{n \to \infty} x_n = \begin{cases} \frac{b}{1-a}, & \text{if } |a| < 1 \text{ or } |a| > 1, x_0 = \frac{b}{1-a}, \\ x_0, & \text{if } a = 1 \text{ and } b = 0, \\ \begin{cases} x_0, & \text{if } n = 0, 2, 4, \ldots \\ b - x_0 & \text{if } n = 1, 3, 5, \ldots \end{cases} \quad a = -1 \\ \infty, & \text{otherwise} \end{cases}$$

Exercise 2. Study the dynamical system generated by the following function

$$f(x) = \begin{cases} ax + b, & \text{if } x \geq 0 \\ b, & \text{if } x < 0 \end{cases}$$

Hint. Draw the graph of the function and compare with the function studied in Example 1.2.

In Example 1.2 since $f(x)$ is linear function, it was easy to solve $f^p(x) = x$.

However, for non-linear functions finding p-periodic points may be, in general, very difficult. Consider the function $f(x) = x^2 - x + 1$ then $f^2(x) = x$ has the form $x^4 - 2x^3 + 2x^2 - x + 1 = 0$. To solve $f^3(x) = x$ we have to solve an equation of order 8. But there is no theory to solve a polynomial equation of order ≥ 5.

Therefore, for a given non-linear function f, it is very important to know for which $p \in \mathbb{N}$ the equation $f^p(x) = x$ has solution (here one interest to the periodic points with prime period p).

This question answered by Sharkovskii's famous theorem [100]. Now we give this theorem. Before, define the following ordering (Sharkovskii ordering) of the set of natural numbers:

$$3 \succ 5 \succ 7 \succ \ldots \quad \text{(all odd numbers, except 1)} \succ$$

$$3 \cdot 2 \succ 5 \cdot 2 \succ 7 \cdot 2 \ldots \quad \text{(all odd numbers multiplied by 2, except 1)} \succ$$

$$3 \cdot 2^2 \succ 5 \cdot 2^2 \succ 7 \cdot 2^2 \succ \ldots \text{(all odd numbers multiplied by } 2^2, \text{ except 1)} \succ$$

$$3 \cdot 2^3 \succ 5 \cdot 2^3 \succ 7 \cdot 2^3 \succ \ldots \text{(all odd numbers multiplied by } 2^3, \text{ except 1)} \succ$$

$$\succ \ldots \succ 2^n \succ \ldots \succ 2^4 \succ 2^3 \succ 2^2 \succ 2 \succ 1.$$

This order consists of the odd numbers in increasing order, 2 times the odds in increasing order, 4 times the odds in increasing order, 8 times the odds etc. at the end the powers of two in decreasing order. Thus, every positive integer appears exactly once somewhere on this order.

The following theorem gives a complete accounting of which periods imply which other periods for continuous function on $\mathbb{R}$.

Theorem 1.2. *(Sharkovskii) Let f be a continuous function on $\mathbb{R}$. If f has a periodic point of least (prime) period p and p precedes q in the above ordering (i.e. $p \succ q$), then f also has a periodic point of least period q.*

Consequently, from this theorem we get

Corollary 1.1.

1. *If a function f has periodic point whose period is not a power of two, then f necessarily has infinitely many periodic points. Conversely, if the function f has only finitely many periodic points, then they all necessarily have periods which are powers of two.*
2. *Since the number 3 is the greatest number in the Sharkovskii's ordering, the existence of the period 3 (by Sharkovskii's theorem) implies the existence of all other periods.*
3. *If a function f has no two periodic points then f does not have any periodic (except fixed) points.*

The converse of Sharkovskii's Theorem is also true. There are functions which have periodic points of period p and no "higher" period points according to the Sharkovskii ordering. Here is an example.

Example 1.3. [33] To produce a continuous function with period 5 and no period 3, consider a function $f : [1,5] \to [1,5]$ as shown in Fig. 1.1, which satisfies

$$f(1) = 3, \ f(3) = 4, \ f(4) = 2, \ f(2) = 5, \ f(5) = 1.$$

This means that the orbit of 1 under function f is 5-periodic:

$$1 \to 3 \to 4 \to 2 \to 5 \to 1.$$

But this function has no any 3-periodic point.

Exercise 3. Give an example of a continuous function which has a period 7 point but does not have period 3 and 5 points.

Hint. Follow the Example 1.3.

Note that the assumption of continuity is important in Sharkovskii's theorem, as the discontinuous function $f(x) = (1 - x)^{-1}$, for which every non-zero value has period 3, would otherwise be a counterexample.

Exercise 4. Prove that for the function $f(x) = (1 - x)^{-1}$, $x \neq 1$ every non-zero x has period 3, i.e. $f^3(x) = x$, $x \neq 0$.

Exercise 5. Find periodic points of the function $f(x) = -(1 - x)^{-1}$.

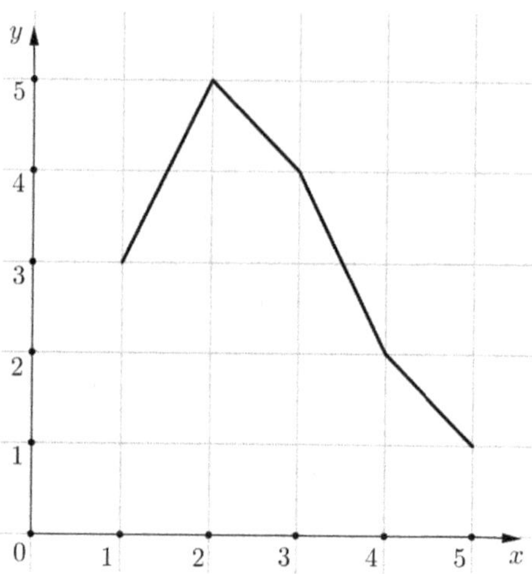

Fig. 1.1   A function with a period 5 point and no period 3 point.

1.1.3   *Multi-dimensional linear systems*

In this subsection we review the investigation steps of multi-dimensional linear systems. We consider $X = \mathbb{R}^m$ and function

$$f : x = (x_1, x_2, \ldots, x_m) \in \mathbb{R}^m \to x' = f(x) = (x'_1, x'_2, \ldots, x'_m) \in \mathbb{R}^m$$

given by

$$x'_k = \sum_{j=1}^{m} a_{kj} x_j + b_k, \quad k = 1, 2, \ldots, m \tag{1.4}$$

where $a_{kj} \in \mathbb{R}$ and $b_k \in \mathbb{R}$.

Introducing the matrix $A = (a_{kj})_{k,j=1}^{m}$ and vector $B = (b_1, \ldots, b_m)$ we write (1.4) as

$$x' = Ax + B. \tag{1.5}$$

Therefore the discrete-time dynamical system is given by the sequence

$$x^{(n+1)} = Ax^{(n)} + B, \quad n \geq 0, \tag{1.6}$$

where $x^{(0)} \in \mathbb{R}^m$ is an initial vector. As before the main problem is to investigate $\lim_{n \to \infty} x^{(n)}$.

Linearity of f allows us to find the following

$$x^{(n)} = A^n x^{(0)} + \sum_{i=0}^{n-1} A^i B, \quad n \geq 0. \tag{1.7}$$

The following proposition is well-known:

Proposition 1.1. *The following are true*

1.
$$\sum_{i=0}^{n-1} A^i = \frac{E - A^n}{E - A}, \quad \text{if} \quad \det(E - A) \neq 0,$$
   where E is the unity matrix.
2. *The operator $Ax + B$ has unique fixed point, denoted by x^*, i.e., $Ax^* + B = x^*$ iff $\det(E - A) \neq 0$. Moreover, the point has the form $x^* = (E - A)^{-1}B$.*
3. *$x^{(n)}$ has the following form*
$$x^{(n)} = A^n(x^{(0)} - (E - A)^{-1}B) + (E - A)^{-1}B.$$

From part 3 of this proposition it follows that to answer the main problem of the dynamical system (1.6) one has to know the behavior of A^n as $n \to \infty$. This problem is solved by diagonalization of matrix A.

Proposition 1.2. [80] *Let $\rho(A)$ denote the spectral radius of A, i.e. the largest absolute value of its eigenvalues. Then $\lim_{n\to\infty} A^n$ exists if and only if $\rho(A) < 1$ or $\rho(A) = 1$, where $\lambda = 1$ is the only eigenvalue on the unit circle, and $\lambda = 1$ is semisimple.*

The following theorem gives characterization of the dynamical system (see Chapter 3 of [36]):

Theorem 1.3. *Let $\det(E - A) \neq 0$ and suppose A has m distinct real eighenvalues $\{\lambda_1, \lambda_2, \ldots, \lambda_m\}$. Then*
$$\lim_{n\to\infty} x^{(n)} = x^* = (E - A)^{-1}B,$$
if and only if for any $j = 1, 2, \ldots, m$ the following holds
$$|\lambda_j| < 1, \quad \text{or} \quad y_j^{(0)} = 0,$$
where $y^{(0)} = Q^{-1}(x^{(0)} - x^*)$, and Q is a nonsingular $m \times m$ matrix whose columns are the eigenvectors, $\{v_1, v_2, \ldots, v_m\}$, of the matrix A.

Remark 1.1. In case when limit of $A^n x$ does not exist, for some $x \in \mathbb{R}^m$, but the set of limit points is finite, then the sequence $A^n x$ is asymptotically periodic, say with a period p. For such a sequence, one can use Proposition 1.2 for the linear mapping A^p, to investigate the limit $\lim_{k\to\infty} A^{pk+i}x$, $i = 0, 1, \ldots, p - 1$. Moreover, if A has an eigenvalue, say λ_j, such that $|\lambda_j| > 1$ then there is a coordinate, say $x_j^{(n)}$, such that $\lim_{n\to\infty} x_j^{(n)} = \infty$.

Exercise 6. In (1.6) take $B \equiv 0$ and matrix A one of the following

$$1) \begin{pmatrix} \alpha & -\beta \\ \beta & \alpha \end{pmatrix}, \beta \neq 0, \quad 2) \begin{pmatrix} \lambda & 1 \\ 0 & \lambda \end{pmatrix}, \quad 3) \begin{pmatrix} \lambda & 0 \\ 0 & \mu \end{pmatrix},$$

$$4) \begin{pmatrix} \alpha & -\beta & 0 \\ \beta & \alpha & 0 \\ 0 & 0 & \lambda \end{pmatrix}, \beta \neq 0, \quad 5) \begin{pmatrix} \lambda & 0 & 0 \\ 0 & \mu & 0 \\ 0 & 0 & \eta \end{pmatrix},$$

$$6) \begin{pmatrix} \lambda & 1 & 0 \\ 0 & \lambda & 0 \\ 0 & 0 & \mu \end{pmatrix}, \quad 7) \begin{pmatrix} \lambda & 1 & 0 \\ 0 & \lambda & 1 \\ 0 & 0 & \lambda \end{pmatrix}$$

and find $\lim_{n \to \infty} x^{(n)}$.

Definition 1.6. Let A_1 and A_2 be matrices and $L_i(x) = A_i x$, $i = 1, 2$ be linear maps of $\mathbb{R}^m$. The maps L_1 and L_2 are called linearly conjugate if there is an invertible matrix B such that the linear map $P(x) = Bx$ satisfies $L_1 = P^{-1} \circ L_2 \circ P$.

It is easy to see that conjugate maps have similar behavior of trajectories.

Remark 1.2. In [33] it is shown that each 2-dimensional and 3-dimensional linear map is conjugate to one of the linear map constructed by a matrix given in Exercise 6.

1.1.4   *Multi-dimensional non-linear systems*

In the previous subsection we saw that the main problem of the theory of dynamical system for linear case can be completely solved. But in non-linear systems the problem is not simple. Such systems mainly reduced to linear systems (linearization) to prove the local and some global results.

Consider a (non-linear) mapping

$$F : x = (x_1, x_2, \ldots, x_m) \in \mathbb{R}^m \to x' = F(x) = (x_1', x_2', \ldots, x_m') \in \mathbb{R}^m$$

given by

$$x_k' = F_i(x_1, x_2, \ldots, x_m), \quad k = 1, 2, \ldots, m, \tag{1.8}$$

where $F_i : \mathbb{R}^m \to \mathbb{R}$, $i = 1, \ldots, m$ is a continuously differentiable single-value function.

As before, the discrete-time dynamical system is given by the sequence

$$x^{(n+1)} = F(x^{(n)}), \quad n \geq 0, \tag{1.9}$$

where $x^{(0)} \in \mathbb{R}^m$ is an initial vector.

Example 1.4. Consider the Hénon map, $H : \mathbb{R}^2 \to \mathbb{R}^2$, defined by

$$H : \begin{cases} x' = H_1(x,y) = 1 + y - ax^2 \\ y' = H_2(x,y) = bx, \quad a, b > 0 \end{cases}$$

which is a quadratic map in dimension two [33]. It is known that for some values of its parameters, the dynamics of the Hénon map is very complex, having infinitely many periodic points. This is one of the most studied examples of dynamical systems that exhibit chaotic behavior.

Now we give linearization of the non-linear mapping F. Assume F has a fixed point $x^* \in \mathbb{R}^m$, i.e., $F(x^*) = x^*$. The function F can be approximated around the fixed point, x^*. Such an approximation can be given by a Taylor expansion of $F_i(x)$, $i = 1, 2, \ldots, m$ around x^*, which has the following form

$$F_i(x) = F_i(x^*) + \sum_{j=1}^{m} \frac{\partial F_i(x^*)}{\partial x_j}(x_j - x_j^*) + \mathcal{R}_m, \tag{1.10}$$

where $\mathcal{R}_m$ is a residual term. Thus, the linearized function around the fixed point is $\hat{F} : \mathbb{R}^m \to \mathbb{R}^m$ of the form $\hat{F}(x) = Ax + B$ where

$$A = J_F(x^*) = \begin{pmatrix} \frac{\partial F_1(x^*)}{\partial x_1} & \frac{\partial F_1(x^*)}{\partial x_2} & \cdots & \frac{\partial F_1(x^*)}{\partial x_m} \\ \frac{\partial F_2(x^*)}{\partial x_1} & \frac{\partial F_2(x^*)}{\partial x_2} & \cdots & \frac{\partial F_2(x^*)}{\partial x_m} \\ \vdots & \vdots & \vdots & \vdots \\ \frac{\partial F_m(x^*)}{\partial x_1} & \frac{\partial F_m(x^*)}{\partial x_2} & \cdots & \frac{\partial F_m(x^*)}{\partial x_m} \end{pmatrix}, \tag{1.11}$$

which is called the Jacobian matrix or Jacobi matrix of $F(x)$ evaluated at x^*, and B is the following constant column vector

$$B = \begin{pmatrix} F_1(x^*) - \sum_{j=1}^{m} \frac{\partial F_1(x^*)}{\partial x_j} x_j^* \\ F_2(x^*) - \sum_{j=1}^{m} \frac{\partial F_2(x^*)}{\partial x_j} x_j^* \\ \vdots \\ F_m(x^*) - \sum_{j=1}^{m} \frac{\partial F_m(x^*)}{\partial x_j} x_j^* \end{pmatrix}. \tag{1.12}$$

This linear function $\hat{F}$ will be used to study the local behavior of the non-linear dynamical system given by the non-linear function F in the proximity of a fixed point x^*, i.e. $\hat{F}$ approximates the non-linear function

F in the neighborhood of this fixed point x^*. Thus, the eigenvalues of the Jacobian matrix J_F determine the local behavior of the nonlinear dynamical system. Below we shall do this point more explicit:

Definition 1.7. A fixed point x^* of a mapping F is called

- hyperbolic point if its Jacobian J_F at x^* has no eigenvalues on the unit circle.
- attracting point if all the eigenvalues of the Jacobi matrix $J_F(x^*)$ are less than 1 in absolute value;
- repelling point if all the eigenvalues of the Jacobi matrix $J_F(x^*)$ are greater than 1 in absolute value;
- a saddle point otherwise.

For a given vector space V over a field $\mathbf{K}$ and $S \subset V$, the span of a set S is the set of all finite linear combinations of elements of S, i.e.,

$$\text{span}(S) = \left\{ \sum_{i=1}^{k} \lambda_i v_i \,\middle|\, k \in \mathbb{N}, v_i \in S, \lambda_i \in \mathbf{K} \right\}.$$

Note that, if S is a finite set, then the span of S is the set of all linear combinations of the elements of S. In the case of infinite S, infinite linear combinations (i.e. where a combination may involve an infinite sum, assuming such sums are defined somehow) are excluded by the definition; a generalization that allows these is not equivalent.

Definition 1.8. Let x^* be a fixed point of a mapping $F : \mathbb{R}^m \to \mathbb{R}^m$.

- The stable eigenspace , $E^s(x^*)$, of the fixed point x^* is
$$E^s(x^*) = \text{span}\{\text{eigenvectors of } J_F(x^*) \text{ whose eigenvalues have modulus} < 1\}.$$
- The unstable eigenspace , $E^u(x^*)$, of the fixed point x^* is
$$E^u(x^*) = \text{span}\{\text{eigenvectors of } J_F(x^*) \text{ whose eigenvalues have modulus} > 1\}.$$
- The center eigenspace , $E^c(x^*)$, of the fixed point x^* is
$$E^c(x^*) = \text{span}\{\text{eigenvectors of } J_F(x^*) \text{ whose eigenvalues have modulus} = 1\}.$$

Remark 1.3.

- The *stable* eigenspace is the geometric locus of all vectors, $x^{(n)}$, which under a sufficient number of forward iterations of the mapping F are mapped in the limit to the fixed point x^*.

- The *unstable* eigenspace is the geometric locus of all vectors, $x^{(n)}$, which under a sufficient number of backward iterations of the mapping F are mapped in the limit to the fixed point x^*.
- The *center* eigenspace is the geometric locus of all vectors, $x^{(n)}$, which are invariant under forward and backward iterations of the mapping F.

The following proposition is clear

Proposition 1.3. *Dimensions of the eigenspaces mentioned in Definition 1.8 satisfy the following*
$$\dim(E^s(x^*)) + \dim(E^u(x^*)) + \dim(E^c(x^*)) = m.$$

A mapping $F : \mathbb{R}^m \to \mathbb{R}^m$ is called a diffeomorphism if it is one to one and F and F^{-1} are continuously differentiable.

A d-dimensional manifold $W \subset \mathbb{R}^m$ is a set·such that for any $x \in W$ there is a neighborhood U for which there exists a diffeomorphism $F : \mathbb{R}^d \to U$ for $d \leq m$.

Definition 1.9. Let x^* be a fixed point of a mapping $F : \mathbb{R}^m \to \mathbb{R}^m$.

- A local stable manifold , $W^s_{loc}(x^*)$, of the fixed point x^* is
$$W^s_{loc}(x^*) = \{x \in U | \lim_{n \to +\infty} F^n(x) = x^* \text{ and } F^n(x) \in U,$$
$$\text{forall } n \in \mathbb{N}\}.$$
- A local unstable manifold , $W^u_{loc}(x^*)$, of the fixed point x^* is
$$W^u_{loc}(x^*) = \{x \in U | \lim_{n \to +\infty} F^{-n}(x) = x^* \text{ and } F^{-n}(x) \in U,$$
$$\text{forall } n \in \mathbb{N}\},$$
where U is a neighborhood of x^*, F^n is the n-th iteration of F, and F^{-n} is the n-th iteration of inverse function F^{-1}.

The following theorem gives the relationship between the stable and unstable eigenspaces and local stable and unstable manifolds in a neighborhood of a fixed point (see [85]):

Theorem 1.4. *Let x^* be a hyperbolic fixed point of a diffeomorphism $F : \mathbb{R}^m \to \mathbb{R}^m$. Then there are $W^s_{loc}(x^*)$ and $W^u_{loc}(x^*)$, that are tangent, respectively to the eigenspaces $E^s(x^*)$ and $E^u(x^*)$ of the Jacobi matrix $J_F(x^*)$. Moreover*

$$\dim W^s_{loc}(x^*) = \dim E^s(x^*), \quad and \quad \dim W^u_{loc}(x^*) = \dim E^u(x^*).$$

The global stable manifold and unstable manifold are defined as the union of the corresponding local manifolds:

- The global stable manifold :

$$W^s(x^*) = \cup_{n=1}^{\infty} \left\{ F^{-n}(W^s_{log}(x^*)) \right\},$$

- The global unstable manifold :

$$W^u(x^*) = \cup_{n=1}^{\infty} \left\{ F^n(W^u_{log}(x^*)) \right\}.$$

A contraction mapping on a metric space (X, ρ) is a function F from X to itself, with the property that there is some nonnegative real number $0 \le \alpha < 1$ such that for all $x, y \in X$,

$$\rho(F(x), F(y)) \le \alpha \, \rho(x, y).$$

For a contraction mapping the main problem of the corresponding dynamical system can be completely solved in light of the following contraction mapping theorem:

Theorem 1.5. *Let (X, ρ) be a non-empty complete metric space with a contraction mapping $F : X \to X$. Then F has a unique fixed-point $x^* \in X$. Moreover, for any initial $x^{(0)} \in X$ the sequence $x^{(n)} = F(x^{(n-1)})$, $n \ge 1$ has the limit $\lim_{n \to \infty} x^{(n)} = x^*$.*

In billiard theory one often uses the Poincaré recurrence theorem, which states that certain systems will, after a sufficiently long but finite time, return to a state very close to the initial state.

To give its formulation we need some definitions: A measure space is a triple $(X, \mathcal{A}, \mu)$, where X is a nonempty set, $\mathcal{A}$ is a σ-algebra[2] on the set X and μ is a measure[3].

Let $T : X \to X$ be a measurable transformation (i.e., if $A \in \mathcal{A}$ then $f^{-1}(A) \in \mathcal{A}$, that is the preimage of any measurable set is measurable) it is called the measure-preserving iff $\forall A \in \mathcal{A}$ one has $\mu(T^{-1}(A)) = \mu(A)$.

Theorem 1.6. *Let T be a measure-preserving transformation of a space with a finite volume. Then for any neighborhood U of any given point there*

---

[2] A σ-algebra is a collection $\mathcal{A}$ of subsets of X that includes the empty subset, is closed under complement, and is closed under countable unions and countable intersections. Any element of the σ-algebra is called measurable set.

[3] A function μ from $\mathcal{A}$ to the extended real number line is called a measure if it satisfies the following properties: $\mu(A) \ge 0$, for all $A \in \mathcal{A}$; $\mu(\varnothing) = 0$; for all countable collections $\{A_i\}_{i=1}^{\infty}$ of pairwise disjoint sets in $\mathcal{A}$ one has

$$\mu\left(\bigcup_{k=1}^{\infty} A_k\right) = \sum_{k=1}^{\infty} \mu(A_k).$$

The measure is a probability measure iff $\mu(X) = 1$.

exists a point $x \in U$ which returns to this neighborhood: $T^n(x) \in U$ for some positive n. The set of points in U that never return to U has zero measure.

Now we give some examples of non-linear dynamical systems.

Example 1.5. ([33], page 219) Consider $F_0 : \mathbb{R}^2 \to \mathbb{R}^2$ given by

$$x' = \frac{x}{2}$$

$$y' = 2y - \frac{15x^3}{8}. \tag{1.13}$$

It is easy to see that $\mathrm{Fix}(F_0) = \{(0,0)\}$. The Jacobi matrix at fixed point is

$$J_F((0,0)) = \begin{pmatrix} 1/2 & 0 \\ 0 & 2 \end{pmatrix}.$$

Thus $(0,0)$ is a saddle point. Consider the following one-dimensional sets

$$l_1 = \{(x,y) \in \mathbb{R}^2 : x = 0\}, \quad l_2 = \{(x,y) \in \mathbb{R}^2 : y = x^3\}.$$

It is clear that $F_0(l_i) \subset l_i$, $i = 1, 2$, i.e. these sets are invariant with respect to F_0. The set l_1 is unstable manifold and the set l_2 is a stable manifold.

Example 1.6. ([95]) Consider $F_1 : \mathbb{R}^2 \to \mathbb{R}^2$ given by

$$x' = \frac{x}{2}$$

$$y' = 2y + x^2. \tag{1.14}$$

It is easy to see that $\mathrm{Fix}(F_1) = \{(0,0)\}$ and $(0,0)$ is a saddle point. Consider the following one-dimensional sets

$$L_1 = \{(x,y) \in \mathbb{R}^2 : x = 0\}, \quad \text{and} \quad L_2 = \{(x,y) \in \mathbb{R}^2 : y = -\frac{4}{7}x^2\}.$$

These sets are invariant with respect to F_1. The set L_1 is an unstable manifold and the set L_2 is a stable manifold.

Example 1.7. The following non-linear dynamical system arises in an investigation of a model of statistical physics. Consider the mapping $F : x \in \mathbb{R}^4_+ \to F(x) = (x'_1, x'_2, x'_3, x'_4) \in \mathbb{R}^4_+$ defined by

$$x'_1 = a \left(bx_1 + b^{-1}x_2 \right)^2$$

$$x'_2 = a^{-1} \left(bx_3 + b^{-1}x_4 \right)^2$$

$$x'_3 = a^{-1} \left(b^{-1}x_1 + bx_2 \right)^2 \tag{1.15}$$

$$x'_4 = a \left(b^{-1}x_3 + bx_4 \right)^2,$$

where $a > 0$, $b > 0$. In Chapter 3 of [90] this dynamical system (1.15) was studied. Its fixed and periodic points and limit points are given depending on parameters a, b.

Denote

$$S^{m-1} = \left\{ x = (x_1, \ldots, x_m) \in \mathbb{R}^m : x_i \geq 0, \ \sum_{i=1}^{m} x_i = 1 \right\}.$$

Exercise 7. Consider mapping $V : S^{m-1} \to S^{m-1}$ given by

     Case $m = 2$:

$$V : \quad \begin{aligned} x_1' &= x_1(1 + ax_2) \\ x_2' &= x_2(1 - ax_1) \end{aligned}, \quad a \in [0, 1].$$

Case $m = 3$:

$$V : \quad \begin{aligned} x_1' &= x_1(1 + x_2 + x_3) \\ x_2' &= x_2(1 - x_1 - x_3) \,. \\ x_3' &= x_3(1 - x_1 + x_2) \end{aligned}$$

Find

$$\lim_{n \to \infty} V^n(x^{(0)}), \quad \text{for any} \ \ x^{(0)} \in S^{m-1}.$$

Hint. Find all fixed points and show that each coordinate of $x^{(n)} = (x_1^{(n)}, \ldots, x_m^{(n)})$ is a monotone and bounded sequence, see [43], [93] and [94] for the theory of such dynamical systems.

1.2   Continuous-time dynamical systems

Continuous-time dynamical systems occur in the solving of problems in which the evolution of the states to be controlled is formulated as a equation (in particular, a differential equation). In this section following [88] we give some auxiliary material devoted to the study of continuous-time dynamical systems.

Remark 1.4. A continuous-time dynamical systems can be given by the Kolmogorov-Chapman equation (which is not a differential equation):

$$M^{[s,t]} = M^{[s,\tau]} M^{[\tau,t]}, \qquad \text{for all} \ \ 0 \leq s < \tau < t,$$

where $M^{[s,t]}$ is an arbitrary (not necessary stochastic) matrix depending on times s, t. This equation gives the time-dependent evolution law of many interacting process (see [21], [73] and the references therein).

Here we will be mostly concerned with continuous-time dynamical systems defined by differential equations. Indeed, some famous examples of dynamical systems can be written in terms of differential equations: the harmonic oscillator, the pendulum and double pendulum, or the N-body problem etc.

Consider the differential equation

$$y'(t) = f(y, t), \quad t \in I = [t_0, t_1],$$
$$y(t_0) = y_0,$$
(1.16)

where[4] $f(x, t)$ is C^r, $r \geq 1$, on some open set $U \subset \mathbb{R}^n \times \mathbb{R}$.

Hence the rate of change, $y'(t)$, depends on the current state $y(t)$ and the time t.

A solution of (1.16) is a map, $y : t \in I \subseteq \mathbb{R} \to y(t) \in \mathbb{R}^n$, which satisfies (1.16), i.e. $y'(t) = f(y(t), t)$, $y(t_0) = y_0$.

The map $y(t)$ has the geometrical interpretation of a curve in $\mathbb{R}^n$, and (1.16) gives the tangent vector at each point of the curve. The main problem of the continuous-time dynamical system is to know the behavior of $y(t)$ as $t \to \pm\infty$.

The following theorem is about the existence and uniqueness of solutions of (1.16):

Theorem 1.7. *There exists a solution of (1.16), denoted by $y(t, t_0, y_0)$ with $y(t_0, t_0, y_0) = y_0$, for sufficiently small neighborhood of t_0. This solution is unique in the sense that any other solution of (1.16) through y_0 at $t = t_0$ must be the same as $y(t, t_0, y_0)$ on their common interval of existence. Moreover, $y(t, t_0, y_0)$ is a C^r function of t, t_0, and y_0.*

Under some weaker assumptions ([17], [54]) on the map $f(x, t)$ one still can obtain existence and uniqueness. Theorem 1.7 only guarantees the existence and uniqueness for sufficiently small time intervals. It is possible to extend in a unique way the time interval of existence of solutions.

Theorem 1.8. *Let $C \subset U \subseteq \mathbb{R} \times \mathbb{R}^n$ be a compact set containing (y_0, t_0). The solution $y(t, t_0, y_0)$ of (1.16) can be uniquely extended backward and forward in t up to the boundary of C.*

Example 1.8. The differential equation $y' = ay$, $y(0) = y_0$ is a very simple model for population growth. This differential equation is obtained by the

---

[4]Recall that f is of class C^r on A if r-th (partial) derivative exists and is continuous at all $x \in A$.

assumption that the rate of growth of the population is proportional to the size of the population [33]. The solution to this equation is $y(t) = y_0 e^{at}$, where $y(0) = y_0$ is the initial population of the species. Hence, if the constant a is positive, $y(t) \to \infty$ as $t \to \infty$ leading to population explosion. If $a < 0$, then $y(t) \to 0$ as $t \to \infty$, leading to extinction.

If the population becomes too large, it may exhaust its resources, and the growth rate may turn negative. This case is considered in the following example.

Example 1.9. Assume that there is some limiting value L for the population, $y(t)$. If it exceeds L, the population should tend to decrease. If $y(t) < L$, there is room for more of the species so $y(t)$ is allowed to increase. The simplest biological model leading to this behavior is

$$y' = ay(L - y), \quad y(0) = y_0. \tag{1.17}$$

Assume that $a > 0$, then we have
1. if $y = L$, then $y' = 0$,
2. if $y > L$ then $y' < 0$,
3. if $y < L$ then $y' > 0$.
We have solution of (1.17) given by

$$y(t) = \frac{L y_0 e^{aLt}}{L - y_0 + y_0 e^{aLt}}.$$

From this formula we get

$$\lim_{t \to +\infty} y(t) = \begin{cases} 0, & \text{if } a < 0 \\ y_0, & \text{if } a = 0 \\ L, & \text{if } a > 0 \end{cases} \tag{1.18}$$

Thus this model conforms more to reality than the exponential growth model (Example 1.8), but (1.18) shows that there is no cyclic behavior or other fluctuations in the population. One might naively expect that the corresponding difference equation behaves similarly. However, following [33] we note that the discrete-time analogue of this example, i.e., $y_{n+1} = ay_n(L - y_n)$, is known as the logistic quadratic map, which has very complex dynamics.

Example 1.10. [88] Consider the equation

$$y' = y^2, \quad y(0) = y_0.$$

This equation has solution

$$y(t, 0, y_0) = -\frac{y_0}{y_0 t - 1},$$

which does not exist for all $t \in \mathbb{R}$, since it becomes infinite at $t = 1/y_0$. Thus the time interval of existence of solution may depend on the initial condition y_0.

Consider a similar example given by the differential equation $y' = 1 + y^2$, with general solution of the form $y(t) = \tan(t + C)$, where C is a constant. It is clear that $y(t)$ goes to infinity in finite time.

Let us now consider differential equations

$$y' = f(y), \quad y \in \mathbb{R}^n, \tag{1.19}$$

where $f(x)$ is a C^r map, with $r \geq 1$, on some open set $U \subset \mathbb{R}^n$. Assume that the solutions $y(t)$ of (1.19) exist for all $t \in \mathbb{R}$.

This solution has properties:

(i) If $y(t)$ is a solution of (1.19), then so is $y(t + \tau)$ for every $\tau \in \mathbb{R}$.

(ii) For any $y_0 \in \mathbb{R}^n$, there exists only one solution of (1.19) passing through this point.

Due to property (i) above, it is clear that it is enough to study solutions of (1.19) with initial conditions of the form $y(0) = y_0$.

Using the solutions $y(t, y_0)$ of (1.19) define a map $\phi : \mathbb{R}^n \times \mathbb{R} \to \mathbb{R}^n$ by

$$\phi(y_0, t) = y(t, y_0).$$

The map ϕ is called a flow and satisfies the following properties:

i) $\frac{d}{dt}\phi(x, t) = f(\phi(x, t))$ for every $t \in \mathbb{R}$ and $x \in \mathbb{R}^n$;

ii) $\phi(x, 0) = x$ for every $x \in \mathbb{R}^n$;

iii) $\phi(\phi(x, s), t) = \phi(x, t + s)$ for every $t, s \in \mathbb{R}$ and $x \in \mathbb{R}^n$;

iv) for fixed t, the map $\phi(x, t)$ is a diffeomorphism of $\mathbb{R}^n$.

It will be convenient to use the notation $\phi^t(x)$ (as in discrete-time iterative case) instead to $\phi(x, t)$ to denote the flow defined by (1.19).

Note that the flow of ϕ determines the evolution of the continuous-time dynamical system defined by (1.19). As mentioned above, the main goal in the theory of dynamical systems is to study the qualitative and geometric properties of solutions to the differential equations, sometimes without going through the process of determining the solution explicitly.

Example 1.11. Consider the system of differential equations

$$u' = v, \quad v' = -\omega^2 u, \quad (u, v) \in \mathbb{R} \times \mathbb{R}, \quad \omega > 0.$$

Then the corresponding flow is the map $\phi : \mathbb{R}^2 \times \mathbb{R} \to \mathbb{R}^2$ given by

$$\phi(x, y, t) = \left(x \cos \omega t + \frac{y}{\omega} \sin \omega t, \quad y \cos \omega t - x\omega \sin \omega t\right).$$

In general, the understanding of the behavior of a given dynamical system is a difficult problem. This problem can be reduced to a more simple one if one knows some invariant sets, i.e., the sets which have the property that if a trajectory starts in the invariant set then it remains in the invariant set, for all of its future, and all of its past.

Definition 1.10. A set $S \subset \mathbb{R}^n$ is called an invariant under the dynamics of (1.19) if for any $y_0 \in S$ we have $\phi^t(y_0) \in S$ for all $t \in \mathbb{R}$. If this condition is satisfied for positive times t then S is called a positive invariant set and, for negative time, as a negative invariant set.

The orbit through y_0, denoted by $O(y_0)$, is the set of points in phase space that lie on a trajectory of (1.19) passing through y_0:

$$O(y_0) = \{\phi^t(y_0) : t \in \mathbb{R}\}.$$

The negative (semi)orbit through y_0 is the set

$$O^-(y_0) = \{\phi^t(y_0) : t \leq 0\}$$

and positive (semi)orbit through y_0 is the set

$$O^+(y_0) = \{\phi^t(y_0) : t \geq 0\}.$$

It is clear that for any $t \in \mathbb{R}$ we have $O(\phi^t(y_0)) = O(y_0)$.

A point $p \in \mathbb{R}^n$ is an equilibrium for the flow of (1.19) if $\phi^t(p) = p$ for all $t \in \mathbb{R}$. Since an equilibrium is a solution that does not change in time, thus providing the most simple example of invariant set.

A point p is called a periodic point of period T for the flow of (1.19) if there exists some number $T > 0$ such that $\phi^T(p) = p$ and $\phi^t(p) \neq p$ for every $0 < t < T$. The orbit $O(p)$ of a periodic point is called a periodic orbit.

The following notations are important in the main problem of the theory of dynamical system.

Definition 1.11. A point q is called an ω-limit (resp. α-limit) point of p for the flow ϕ^t if there exists a sequence of integers $\{t_k\}_{k \in \mathbb{N}}$ going to $+\infty$ (resp. $-\infty$) as k goes to infinity such that

$$\lim_{k \to \infty} \phi^{t_k}(p) = q.$$

The set of all ω-limit (resp. α-limit) points of p for ϕ^t is called the ω-limit set of p and is denoted by $\omega(p)$ (resp. $\alpha(p)$).

Theorem 1.9. *The following assertions hold*

- *The limit sets $\omega(p)$ and $\alpha(p)$ are closed and invariant.*
- *If $O^+(p)$ (resp. $O^-(p)$) is contained in some compact subset of $\mathbb{R}^n$, then $\omega(p)$ (resp. $\alpha(p)$) is nonempty, compact, and connected.*

Example 1.12. Consider differential equation

$$u' = v, \quad v' = -u, \quad (u, v) \in \mathbb{R} \times \mathbb{R}. \tag{1.20}$$

The solution passing through the point $(u, v) = (1, 0)$ at $t = 0$ is given by

$$(u(t), v(t)) = (\cos t, -\sin t).$$

The integral curve passing through $(1, 0)$ at $t = 0$ is the set

$$\{(u, v, t) \in \mathbb{R}^3 : (u(t), v(t)) = (\cos t, -\sin t), \ t \in \mathbb{R}\}.$$

An equilibrium of (1.20) is the point $(0, 0)$. Every other point in $\mathbb{R}^2$ is a periodic point for (1.20). The (periodic) orbit passing through $(u, v) = (1, 0)$ is the unit circle $u^2 + v^2 = 1$.

Example 1.13. For the logistic model for the growth of a population

$$y' = ay(1 - y), y \in \mathbb{R}, \quad a > 0$$

we have

- $\omega(0) = \alpha(0) = \{0\}$ and $\omega(1) = \alpha(1) = \{1\}$;
- if $y_0 \in (0, 1)$, then $\alpha(y_0) = \{0\}$ and $\omega(y_0) = \{1\}$;
- if $y_0 > 1$, then $\alpha(y_0) = \emptyset$ and $\omega(y_0) = \{1\}$;
- if $y_0 < 0$, then $\alpha(y_0) = \{0\}$ and $\omega(y_0) = \emptyset$.

Example 1.14. Consider the following homogeneous linear differential equation of the first order

$$u' = 3u - 4v, \quad v' = 4u - 7v.$$

To solve this system of differential equations, at some point of the solution process we shall need a set of two initial values. In this case, let us pick $u(0) = v(0) = 1$.

General solutions of this equation are

$$u(t) = 2Ae^t + Be^{-5t}, \quad v(t) = Ae^t + 2Be^{-5t}.$$

Using the initial condition $u(0) = v(0) = 1$, when $t = 0$, we get

$$u(t) = \frac{2}{3}e^t + \frac{1}{3}e^{-5t}, \quad v(t) = \frac{1}{3}e^t + \frac{2}{3}e^{-5t}.$$

Thus if $u(0) = x$, $v(0) = y$ then

$$\phi^t(x,y) = \left(\frac{4x-2y}{3}e^t + \frac{2y-x}{3}e^{-5t}, \ \frac{2x-y}{3}e^t + \frac{4y-2x}{3}e^{-5t} \right).$$

Consequently,

$$\lim_{t \to -\infty} \phi^t(x,y) = \begin{cases} (0,0), & \text{if } x = 2y \\ (\infty,\infty), & \text{if } x \neq 2y. \end{cases}$$

$$\lim_{t \to +\infty} \phi^t(x,y) = \begin{cases} (0,0), & \text{if } 2x = y \\ (\infty,\infty), & \text{if } 2x \neq y. \end{cases}$$

1.3   Definitions and problems of billiards

1.3.1   *Definitions*

Billiard sports, are a wide variety of games of skill generally played with a cue stick, which is used to strike billiard balls and thereby cause them to move around a cloth-covered rectangular billiards table bounded by elastic bumpers known as cushions.

One ball is then struck with the end of a "cue" stick, causing it to bounce into other balls and reflect off the sides of the table. Real billiards can involve spinning the ball so that it does not travel in a straight line, but the mathematical study of billiards generally consists of reflections in which the reflection and incidence angles are the same. However, strange table shapes such as circles, ellipses and anther figure can be considered (see Fig.1.2).

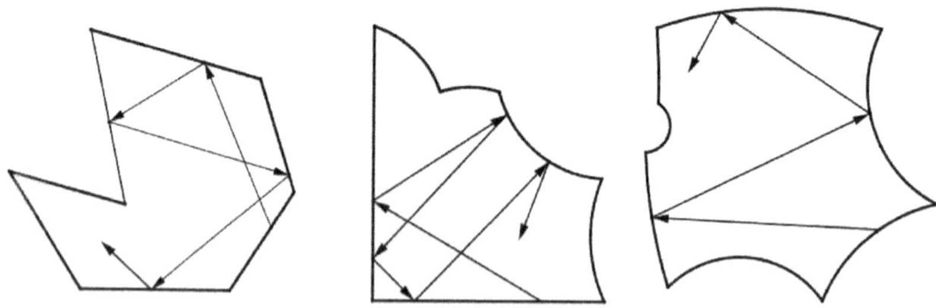

Fig. 1.2   Examples of billiard tables and trajectories.

Let us first consider smooth billiard tables on 2-dimensional euclidean space.

Definition 1.12. A domain $D \subset \mathbb{R}^2$ is called a billiard table if D satisfies the following conditions:

(1) is an open, bounded, and connected domain;
(2) its boundary $\Gamma = \partial D$ is a finite union of smooth compact curves.

Let the moving ball in the billiard table $D \subset \mathbb{R}^2$ has position $q \in D$ and velocity vector $v \in \mathbb{R}^2$.

Then the curves of the boundary of the billiard table and the velocity vector of the moving ball satisfy the following conditions:

1. The curves are disjoint but may have common endpoints.
2. At point $q \in D$ the billiard travels in a straight line parallel to the direction of the velocity vector at point q until it hits the boundary Γ. Thus the billiard always moves in a straight line.
3. Define the billiard trajectory as the segment $\overline{p_1 p_2}$, where p_1 and p_2 are the points on Γ, where the billiard consecutively hits the boundary.
4. Define $n(p)$ as the inward pointing normal vector at point $p \in \Gamma$.
5. Consider $p_1, p_2,$ and p_3 as three consecutive points the billiard contacts with the boundary of the billiard table. At point p_2 define the angle of incidence as the angle between the inward pointing normal vector $n(p_2)$ at point p_2 and the billiard trajectory $\overline{p_1 p_2}$. Similarly, define the angle of reflection as the angle between $n(p_2)$ and the billiard trajectory $\overline{p_2 p_3}$.
6. At every point $p \in \Gamma$, where the billiard hits the boundary, the angle of incidence is the same as the angle of reflection. This is an empirical fact in physics.

Definition 1.13. If the billiard is given in a domain $D \subset \mathbb{R}^2$ with a bounded and closed boundary $\Gamma = \partial D$ then a trajectory of the mathematical billiard is a polygonal chain (i.e. a connected series of line segments) entered in the curve Γ, this trajectory can be unambiguously constructed by the initial segment (see Fig.1.2).

Many interesting problems can arise in the detailed study of billiards trajectories.

In multi-dimensional case Definition 1.13 can be generalized as

Definition 1.14. A billiard trajectory in a smooth convex body $D \subset \mathbb{R}^d$ is a polygon $P \subset D$, with all its vertices on the boundary of D, and at each vertex the direction of line changes according to the specular reflection rule.

The trajectory with starting point (p, v), (where p is point where the ball is placed, and v is its speed) is called periodic, if after some time (after the period) the point comes back to the initial position (p, v).

Definition 1.15. Let P be a periodic billiard trajectory, i.e., a closed polygon. Its number of vertices is called its length.

1.3.2   *The billiard as a two-dimensional non-linear dynamical system*

In this subsection we give an example (see [83]) of billiard which is a two-dimensional non-linear dynamical system. Consider a bounded, strictly convex region D in the plane $\mathbb{R}^2$, with closed γ-boundary curve. The billiard ball moves along straight lines inside and is reflected at the boundary γ under equal angles. In general, the trajectory of this motion can be very complicated for most regions D.

Now we introduce a mapping dynamical systems of which is a mathematical billiard. Recall that a positively oriented curve is a planar simple closed curve such that when traveling on it one always has the curve interior to the left (and consequently, the curve exterior to the right). Consider the positive orientation on γ. The mapping will take a given oriented line segment joining two points on the boundary of D into the one obtained by reflection at its end point. Let s be a parameter along the oriented boundary γ of D, which is proportional to the arc length, the factor being so chosen that one revolution corresponds to 2π. Denote by α the angle which a line segment forms with the positively oriented tangent to the boundary at the initial point of the segment (see Fig. 1.3). Under these notations, the initial states of the billiard form an open strip

$$\mathcal{R} = \{(s, \alpha) : \ 0 < \alpha < \pi\}.$$

On this set define the mapping $F : (s, \alpha) \to (s', \alpha')$ as

$$F : \quad \begin{cases} s' = s + f(s, \alpha), \\ \alpha' = g(s, \alpha), \end{cases} \tag{1.21}$$

where s', α' correspond to the coordinates of the reflected line segment issuing from the endpoint (see Fig. 1.3).

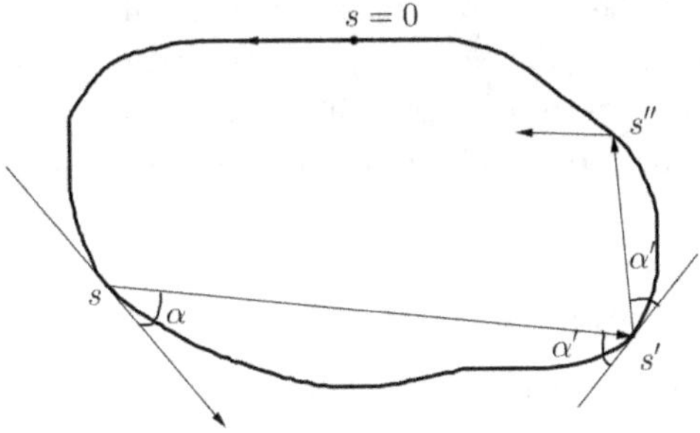

Fig. 1.3

Note that in mapping F the function $f(s, \alpha)$ is defined only up to an integer multiple of 2π which can be fixed as follows. The mapping (1.21) can be extended to a homeomorphism[5] of $\mathcal{R}$. As the angle α goes to zero, $\alpha \to 0$, the line segments from (s, α) to (s', α') become shorter and consequently one can set

$$f(s, 0) = 0 \quad \text{and} \quad g(s, 0) = 0.$$

This choice fixes $f(s, \alpha)$. Moreover, we note that for $\alpha \to \pi$ one has also $f(s, \alpha)$ tending towards an integer multiple of 2π, but this integer is not zero. If the orientation of the curve is chosen as above, keeping s fixed and letting α increase from 0 to π one has

$$f(s, \pi) = 2\pi \quad \text{and} \quad g(s, \pi) = \pi.$$

Then the image print (s', α') will travel once around the boundary of length 2π.

Theorem 1.10. *On a strictly convex billiard table D there exist infinitely many distinct periodic orbits.*

---

[5]Recall the following definitions: let A, B be some sets, and $f : A \to B$ a function then

- $f(x)$ is called one-to-one if $f(x) \neq f(y)$ whenever $x \neq y$.

- The function is called onto if for any y in B there is an $x \in A$ such that $f(x) = y$.

- The function f is called a homeomorphism if it is one-to-one, onto, and continuous, and $f^{-1}(x)$ is also continuous.

- The function f is a C^r-diffeomorphism if f is a C^r-homeomorphism such that $f^{-1}(x)$ is also C^r.

Proof. Note that the set $\mathcal{R}$ becomes an annulus if (s, α) is identified with (s', α) for $(s' - s)/2\pi \in \mathbb{Z}$. Now to prove theorem one uses the mapping (1.21) which is an area-preserving annulus mapping whose periodic points correspond to the periodic orbits of the billiard (see [83] for more details).

$\square$

1.3.3 *The main problem*

As mentioned above a mathematical billiard is an example of a dynamical system, that is a system that evolves in time. Usually in consideration of a dynamical system one is interested in determining the asymptotic behavior, or long-time evolution of the trajectories. Similarly, the main mathematical problem of billiards consists in describing possible types of billiard trajectories in the domain D.

The simplest principle to solve this problem is the description of periodic (cyclic) trajectories and acyclic trajectories.

The considered problem concerning periodic trajectories, in particular, reduces to a question of an existence: whether in any domain D exist periodic trajectories? Moreover, an interesting problem is to find possible lengths of period for billiards in a given domain. Anther question is about criterion of periodicity: how to recognize whether there will be the corresponding trajectory periodic by given starting point (p, v)? Moreover one can ask the following questions: Are trajectories dense? If a trajectory is dense, is it equidistributed?

The interesting problem of existence of periodic trajectories on billiard tables is considered since 1775 when Fagnano discovered the fact that every acute triangle admits such a trajectory (see [31] and references therein). This trajectory is called the Fagnano trajectory, which has a very simple geometric description as being the pedal curve of the orthocenter of the triangular table, i.e. the polygonal line joining successively the projections of the orthocenter on the sides of the table (the so-called orthic triangle) see Fig. 1.4. In [77] it was proved that every rational polygon (i.e. a polygon whose angles between any two of its sides are rational multiples of π) has periodic billiard trajectories.

The paper [31] gives a characterization of the family of the convex polygons that are pedal strong Fagnano trajectory in some convex polygonal billiard table. It is shown that this family of polygons is the set of all convex Poncelet n-gons and that a convex cyclic polygonal billiard table has a pedal strong Fagnano trajectory if and only if it is a contact polygon of a convex Poncelet polygon. We will give these results in Chapter 3.

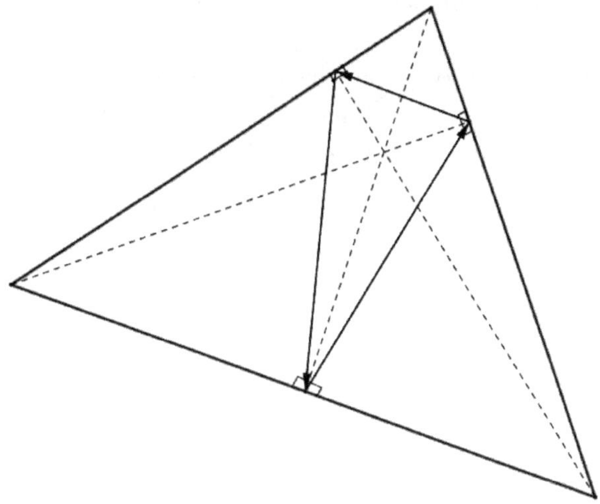

Fig. 1.4   The Fagnano trajectory.

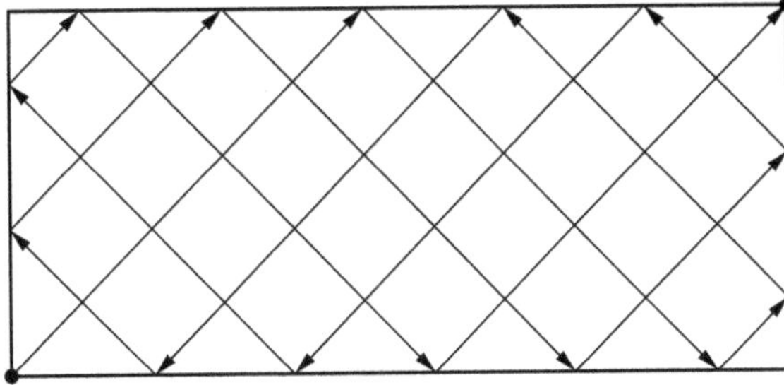

Fig. 1.5   The trajectory started from a corner and finished in another corner.

In [109] and [115] some Fagnano trajectories for polygonal dual billiards are considered.

The following question, is very simple to formulate, but has been long open:

Is there a periodic trajectory in an obtuse triangle?

Note that the billiards on a rectangular table have simple mathematical rules. If we assume that the table has no friction, then the ball will go on forever, unless it eventually hits one of the corners of the table. If it hits a corner, then the ball will stop moving (see Fig. 1.5).

The next chapters are devoted to a theory of billiards and their applications.

Bibliographical notes. This chapter is helpful for the beginners of the theory of dynamical systems (in particular, mathematical billiards). This chapter is based on [33], [35], [36], [43], [54], [65], [83], [88], [93], [94] and many internet sources.

Chapter 2

Billiards in elementary mathematics

In this chapter we consider some examples of billiards and their applications to problems of elementary mathematics. The main aim of this chapter is to show that the theory of mathematical billiards is very interesting and to illustrate it as a very useful theory for solving many difficult problems of elementary mathematics.

2.1 Pouring problems

The following problem can be solved straightforwardly by using a billiard considered on a parallelogram table.

Problem 2.1. *There are two vessels with capacities 7 and 11 liters and there is a greater of a flank filled by water. Using these vessels how to measure exactly 1 liter of water?*

Solution. In the problem the billiard table can be considered as a parallelogram (see Fig. 2.1).

The sides of the table must be 7 and 11. Following the trajectory showed in Fig. 2.1 we can conclude the following:

1. The ball starts its trajectory at the point $(0,0)$ (the left bottom vertex). This position of the ball means that both vessels are empty.
2. In the next position it goes to $(0,11)$, which means the big vessel is full and the small vessel is empty.
3. Then it goes to the position $(7,4)$, which means that water has been poured from the large vessel to the small one.
4. The next position $(0,4)$ corresponds to the act that the small vessel has been poured out.

29

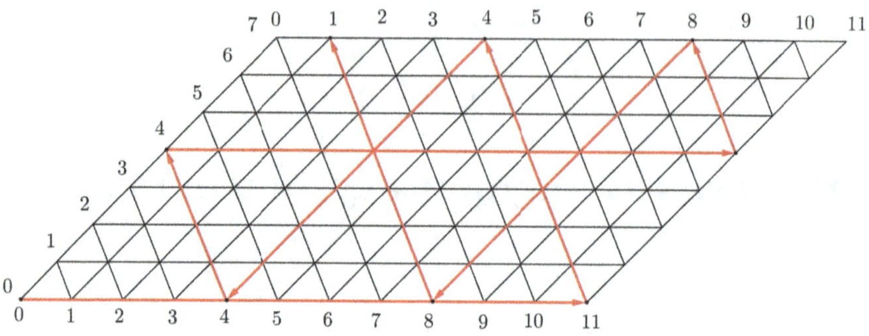

Fig. 2.1   The trajectory separating 1 liter of water.

We should continue to follow the trajectory until one of the vessels will contain exactly 1 liter of water. The Fig. 2.1 shows that in the 8th step the large vessel contains exactly 1 liter. Then the described algorithm gives the solution of the problem.

Remark. If one first directs the ball to point $(7,0)$ (the left top vertex) then to get 1 liter, one has to do 25 steps. It is easy to check that by the mathematical billiard of Fig. 2.1 one can measure i liter of water for any $i = 1, 2, \ldots, 11$. Just continue the trajectory until to a point with a coordinate equal to i.

Problem 2.2. *There is a vessel with capacity 8, which is full of water. There are two empty vessels with capacities 3 and 5 liters. How to pour the water in two greater vessels equally (i.e. both vessels must contain exactly 4 liter of water)?*

Solution. The table for this problem is a 3×5 parallelogram (see Fig. 2.2).

The large diagonal of the parallelogram, which corresponds to the vessel with capacity 8, is divided to 8 parts by the inclined straight lines. Following the trajectory shown in Fig. 2.2, we should continue until it separates 4 liters. The trajectory is

$$(0,0,8) \to (0,5,3) \to (3,2,3) \to (0,2,6) \to$$

$$(2,0,6) \to (2,5,1) \to (3,4,1) \to (0,4,4).$$

This trajectory gives the algorithm of the solution.

Remark. If two smaller vessels have coprime (relatively prime) capacities (i.e. the capacity (volume) numbers have not a common divisor

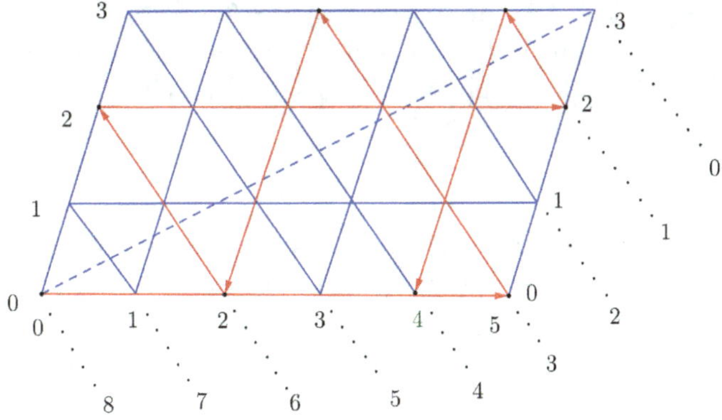

Fig. 2.2   The trajectory dividing 8 liters to two 4 liters.

$\neq 1$) and the biggest vessel has a capacity larger (or equal) than sum of the capacities of the smaller vessels, then using these three vessels one can measure water with liters from 1 up to the capacity of the mid vessel. For example, if there are three vessels with capacities 12, 13 and 26 respectively, then one can measure l liter of water for any $l \in \{1, 2, ..., 13\}$.

Exercise 8. a) There are two vessels with capacities 6 and 9 liters and there is a greater of a flank filled by water. Find all possible volumes of water which can be measured by these vessels.

b) There is a vessel with capacity 10, which is full of water. There are two empty vessels with capacities 4 and 6 liters. Pour the water in two greater vessels equally.

2.2   Billiard in the circle

The circle enjoys rotational symmetry, and a billiard trajectory is completely determined by the angle θ made with the circle. This angle remains the same after each reflection. Namely, define a mapping $T_\theta : [0, 1] \to [0, 1]$ such that $T_\theta(x) = x + \theta (mod\,1)$ where $x \in [0, 1]$ and $\theta \in \mathbb{R}$. Here x denotes the initial position and θ denotes the angle of rotation along the circle. It is easy to see that the n iterations of the map T_θ can be given as follows:

$$T_\theta^n(x) = x + n\theta (mod\,1). \tag{2.1}$$

Exercise 9. Prove the formula (2.1).

For given $x^{(0)} \in [0,1]$ the sequence $x^{(n)} = T_\theta(x^{(n-1)})$, $n \geq 1$ is called an orbit of the mapping T_θ. In some literature this sequence is called trajectory of $x^{(0)}$ under mapping T_θ, and some times it is called a discrete time dynamical system [33], [35].

Theorem 2.1.

1. *If $\theta \in \mathbb{R}$ is a rational number, then every orbit of the mapping $T_\theta(x)$ is periodic. Moreover, if $\theta = p/q$ then the length of periodicity is equal to q.*

2. *If $\theta \in \mathbb{R}$ is an irrational number, then every orbit of the mapping $T_\theta(x)$ is dense in $[0,1]$. In other words, every interval of the circle contains points of this orbit.*

Proof.   1. Let θ be a rational number, say $\theta = p/q$ then, for any $x \in [0,1]$, from (2.1) it follows that if $T^n(x) = x$ for some $n \in \mathbb{N}$ then $x + n\theta \,(mod\,1) = x$. From the last equality it follows that $n\theta = \frac{np}{q} \in \mathbb{Z}$, i.e., $n = qk$ for some $k = 1, 2, \dots$. This completes proof of part 1.

2. We shall show that each $T_\theta^n(x)$ is distinct. For two distinct natural numbers i and j assume $T_\theta^i(x) = T_\theta^j(x) + k$ for some positive integer k. That is $x + i\theta = x + j\theta + k$. So, $\theta(i - j)$ is a natural number. Since θ is irrational, $i = j$, a contradiction.

Now divide the interval $[0,1]$ into n intervals, each with length $1/n$. By the fact showed above, we know that the $n + 1$ terms $x, T_\theta(x), \dots, T_\theta^n(x)$ are distinct. Then by the pigeonhole principle[1] there exists an interval that includes at least two of the terms $T_\theta^i(x)$ and $T_\theta^j(x)$ for $i \neq j$. Consequently, for every positive integer n there exist two distinct natural numbers i and j such that

$$|T_\theta^i(x) - T_\theta^j(x)| < \frac{1}{n}.$$

Let us now prove that every neighborhood of an arbitrary point $y \in [0,1]$ contains the term $T_\theta^m(x)$. Without loss of generality we can show that every neighborhood of the initial point x contains the term $T_\theta^m(x)$. It is clear that the euclidean distance between two points of the interval is invariant under iterations of the map T_θ, therefore, for natural numbers i, j, and n we have

$$d(T_\theta^{i-j}(x), x) = d(T_\theta^j(T_\theta^{i-j}(x)), T_\theta^j(x)) = d(T_\theta^i(x), T_\theta^j(x)) < \frac{1}{n}.$$

Thus, every neighborhood of the initial point x contains the term $T_\theta^{i-j}(x)$. Hence, every orbit of T_θ is dense in $[0,1]$. $\qquad\square$

---

[1] The pigeonhole principle states that if n items are put into m containers, with $n > m$, then at least one container must contain more than one item.

Corollary 2.1. *If θ is irrational, then the T_θ-orbit has infinitely many points in any arc Δ of the circle.*

Let us study the sequence $x_n = x + n\theta \,(mod\,1)$ with an irrational θ. If $\theta = \frac{p}{q}$, this sequence consists of q elements which are distributed in the circle very regularly. Should one expect a similar regular distribution for an irrational θ?

The adequate notion is that of *equidistribution* (or *uniform* distribution). Given an arc Δ, denote $|\Delta|$ to be the length of the arc. Let $k(n)$ be the number of terms in the sequence $x_0, \ldots, x_{n-1}$ that lie in Δ. The sequence is called equidistributed on the circle if

$$\lim_{n \to \infty} \frac{k(n)}{n} = |\Delta|,$$

for every Δ.

The next theorem is due to Kronecker and Weyl, which is a special case of the ergodic theorem [12].

Theorem 2.2. *If θ is irrational, then the sequence $x_n = x + n\theta \,(mod\,1)$ is equidistributed on the circle.*

The following theorem gives a characterization of a billiard on a circular table:

Theorem 2.3. *Any billiard trajectory in a circle never comes in some concentric circle in boundary of which all segments of the trajectory are tangent lines (see Fig. 2.3).*

Proof. Consider a billiard trajectory $P_0 P_1 P_2 \ldots$ (a polygon with vertices $P_0, P_1, P_2, \ldots$). By definition of the trajectory it follows that

$$\overline{P_0 P_1} = \overline{P_1 P_2} = \overline{P_2 P_3} = \ldots$$

Note that for any $k = 1, 2, 3\ldots$ the triangles $P_{k-1}OP_k$ and P_kOP_{k+1} are equal, because they are isosceles with equal corners at the bases. Consequently, the following angles are equal (see Fig. 2.3):

$$\angle P_0 O P_1 = \angle P_1 O P_2 = \angle P_2 O P_3 = \ldots$$

Furthermore, it is easy to see that the middle points $\{K_i\}$ of all segments of trajectories are at an identical distance from the center of the circle, thus, they are located on a circle with the same center O. Therefore, each billiard trajectory is located in a circular ring (see Fig. 2.3). $\square$

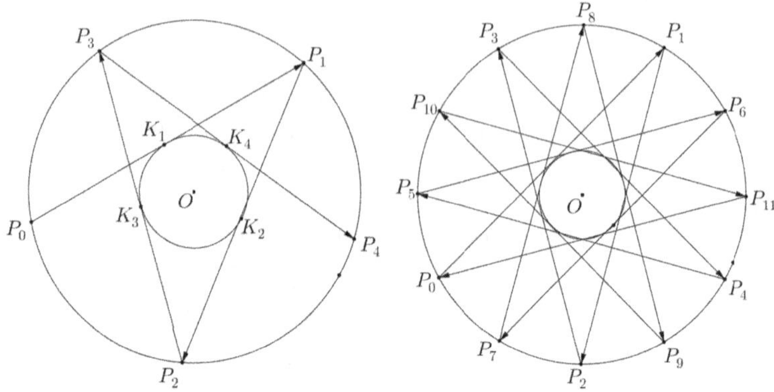

Fig. 2.3   The trajectory never comes inside of inner circle.

Remark 2.1. From Theorem 2.3, it follows that any billiard in a circle is not ergodic, here the ergodicity, in particular, means passing (at some time) the billiard ball through any small circle of a domain Q (this is called everywhere density of the billiard trajectory). Therefore the billiard trajectory in a circle is not everywhere dense.

Problem 2.3. *The billiard ball is on a circle table Γ of radius $R = \frac{1}{2\pi}$ at distance l from its center. Under which conditions on parameters l and φ, the billiard trajectory is periodic? Find the area which is everywhere dense sweeps up, if the ball is released at an angle φ to the diameter on which it is (see Fig. 2.4).*

Solution. Using the law of sines, for the billiard trajectory, one can find the rotation angle: $\theta = \arccos(2\pi l \sin \varphi)$. Therefore, by Theorem 2.1 we have that the trajectory is periodical if the angle θ is rational. If this θ is an irrational number, then the trajectory everywhere dense fills a ring with an internal circle γ of radius equal to $l \sin \varphi$ (see Fig. 2.5).

Exercise 10. Show that any non-periodic trajectory on a circle table does not contain parallel line segments.

Exercise 11. Prove that if a periodic trajectory on a circle table contains two parallel line segments then the trajectory has an even number of line segments.

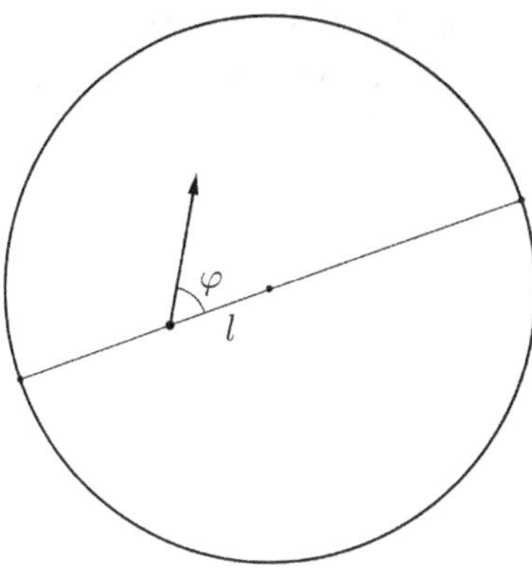

Fig. 2.4   The initial position of the ball.

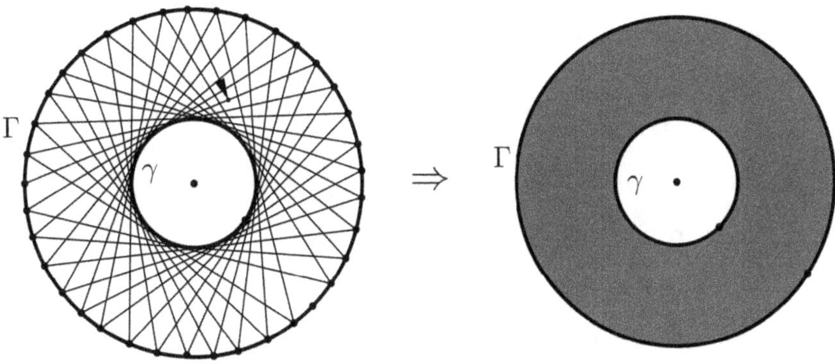

Fig. 2.5   The dense ring.

Exercise 12.

1. In a circular billiard table n billiard balls are located. Prove that they can be shot in such directions that
   a) any two of them would never have collision;
   b) any two of them will have a collision.
2. On the circular billiard table of radius 1 there are two balls with radius 0.0001 each. Prove that they can be shot in such directions that they will surely meet (collision) with each other.

3. Is it possible to shoot two balls of radius 10^{-10} in circular billiard tables of radius 1 so that they are never faced?

2.3   Application of billiards in problems of mathematical olympiad

Now we shall give some applications of Theorems 2.1 and 2.2 in solving some problems of mathematical olympiad.

2.3.1   *Circular table*

Problem 2.4. *Distribution of first digits. Consider the sequence*

$$1, 2, 4, 8, 16, 32, 64, 128, 256, 512, 1024, \ldots$$

consisting of consecutive powers of 2.

Can a power of 2 start with 2018?

Is a term in this sequence more likely to start with 3 or 4?

Solution. Let us consider the second question: 2^n has the first digit k if, for some non-negative integer q, one has

$$k10^q \leq 2^n < (k+1)10^q.$$

Take logarithm base 10:

$$\log k + q \leq n \log 2 < \log(k+1) + q. \tag{2.2}$$

Since q is of no concern to us, let us consider fractional parts of the numbers involved. Denote by $\{x\}$ the fractional part of the real number x. Inequalities (2.2) mean that $\{n \log 2\}$ belongs to the interval $I = [\log k, \log(k+1))$. Note that $\log 2$ is an irrational number. Thus by Theorem 2.1 there is a number n_0 such that $2^{n_0} = k \ldots$. Using Theorem 2.2, we obtain the following result.

Corollary 2.2. *The probability $p(k)$ for a power of 2 to start with digit k equals $\log(k+1) - \log k$.*

The values of these probabilities are approximate as follows:

$$p(1) = 0.301, \quad p(2) = 0.176, \quad p(3) = 0.125, \quad p(4) = 0.097,$$

$$p(5) = 0.079, \quad p(6) = 0.067, \quad p(7) = 0.058, \quad p(8) = 0.051, \quad p(9) = 0.046.$$

We see that $p(k)$ monotonically decreases with k; in particular, 1 is about 6 times as likely to be the first digit as 9.

Exercise 13. a) What is the distribution of the first digits in the sequence $2^n C$ where C is a constant?

b) Find the probability that the first m digits of a power of 2 is a given combination $k_1 k_2 ... k_m$.

c) Investigate similar questions for powers of other numbers.

d) Prove that if p is such that $p \neq 10^q$ (for some $q = 1, 2, ...$) then the sequence $p, p^2, p^3, ...$ has a term where the first m digits is a given combination $k_1 k_2 ... k_m$.

Remark 2.2. Surprisingly, many real life sequences enjoy a similar distribution of first digits! This was first noted in 1881 in a 2-page article by American astronomer S. Newcomb. This article opens as follows: That the ten digits do not occur with equal frequency must be evident to any one making much use of logarithmic tables, and noticing how much faster the first pages wear out than the last ones. The first significant figure is oftener 1 than any other digit, and the frequency diminishes up to 9.

Problem 2.5. *Is there a natural number n such that $0 < \sin n < 10^{-2018}$?*

Solution. The answer is "Yes"! In order to prove this, consider a billiard on a circle with radius 1, which corresponds to the rotation number $\theta = 1$ radian (see Fig. 2.6). Then sequence $\sin 0, \sin 1, \sin 2, ...$ on $[-1, 1]$ corresponds to the trajectory $0, 1, 2, ...$ of the billiard with the starting point 0. Since 1 radian is irrational, by Theorem 2.1 we get the result. Note that the question is trivial if one considers $x \in \mathbb{R}$ instate of $n = 1, 2, ...$

Problem 2.6. *Prove that for any $a \in [-1, 1]$ there is a sequence $\{n_k\}_{k=1}^{\infty}$ of natural numbers such that*

$$\lim_{k \to \infty} \sin n_k = a,$$

i.e., the set of all limit points of the sequence $\{\sin n\}_{n=1}^{\infty}$ is $[-1, 1]$.

Proof. One can realize the proof by using the billiard considered in the solution of Problem 2.5 and Corollary 2.1. $\qquad\square$

Exercise 14. Solve Problems 2.5 and 2.6 replacing sin by cos.

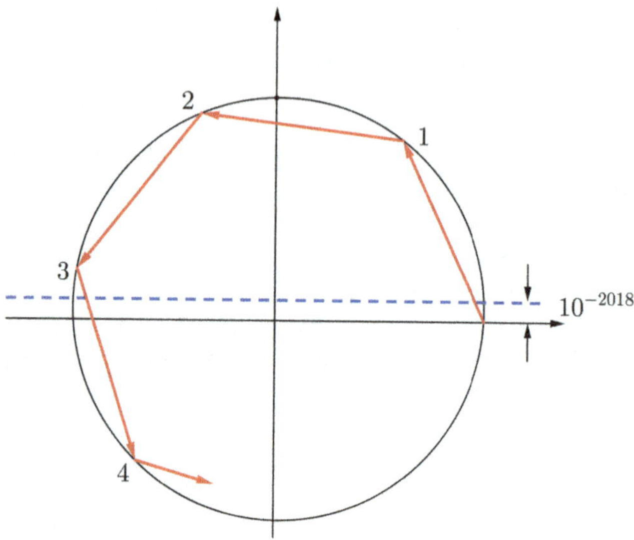

Fig. 2.6   Billiard in the circle.

Problem 2.7. *Let $y_k = ak + b \,(\mathrm{mod}\, 1)$, $k \in \mathbb{Z}$. For any $b \in \mathbb{R}$, prove that*

   1. If a is a rational number then $\{y_k\}_{k \in \mathbb{Z}}$ is a finite subset of $[0, 1)$.
   2. If a is an irrational number then $\{y_k\}_{k \in \mathbb{Z}}$ is a dense subset on $[0, 1)$.

Proof. Consider $[0, 1)$ as a unit circle S^1, stick together 0 and 1. Now let's reel up the numerical axis $\mathbb{R}$ on the circle S^1 in such a way that b coincides with 0 (and 1) (see Fig. 2.7). Then each points $ak + b$, $k \in \mathbb{Z}$ will be replaced on S^1 and we obtain their fractional part. These points on S^1 then can be considered as a billiard trajectory

$$y_0 = b, y_1 = y_0 + a\,(\mathrm{mod}\, 1), y_2 = y_0 + 2a\,(\mathrm{mod}\, 1), \ldots$$

Therefore by Theorem 2.1 one completes the proof. $\square$

2.3.2   *Motion on torus*

In this section, we give some properties of the motion on a torus. This will be used to solve some difficult mathematical (olympiad) problems.

Definition 2.1. A torus is a surface generated by revolving a circle in three-dimensional space, $\mathbb{R}^3$, about an axis coplanar with the circle (see Fig. 2.8).[2]

---

[2]https://en.wikipedia.org/wiki/Torus

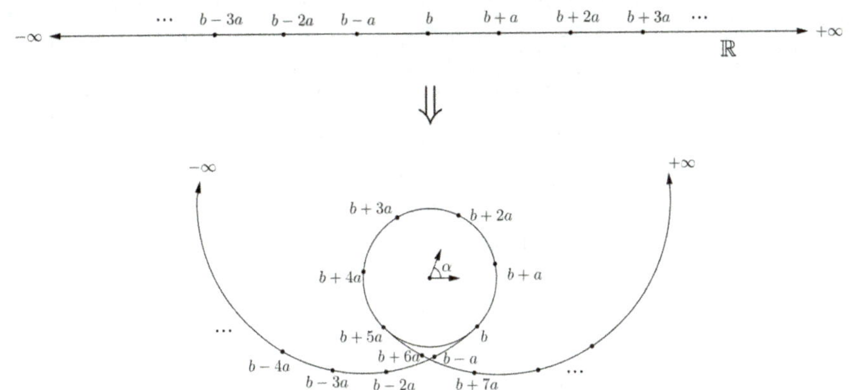

Fig. 2.7   Reeling of $\mathbb{R}$ on the circle S^1.

(a)   *(b)*

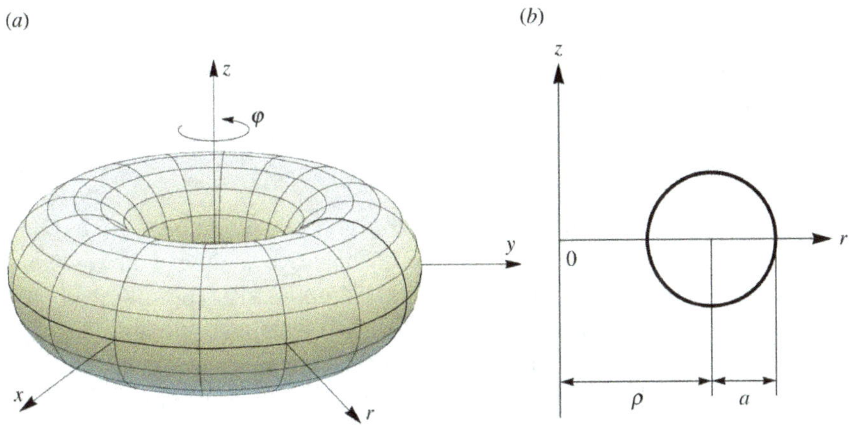

Fig. 2.8   The torus (a) obtained by revolving the circle shown in (b).

A torus should not be confused with a solid torus, which is formed by rotating a disc, rather than a circle around an axis.

Note that a torus can be considered as the Cartesian product of two circles: $S^1 \times S^1$.

A torus can be defined parametrically with two parameters $\theta, \varphi \in [0, 2\pi]$ by:

$$\begin{cases} x(\theta, \varphi) = (\rho + a\cos\theta)\cos\varphi \\ y(\theta, \varphi) = (\rho + a\cos\theta)\sin\varphi \\ z(\theta, \varphi) = a\sin\theta, \end{cases} \tag{2.3}$$

where θ, φ are angles which make a full circle, so that their values start and end at the same point,

ρ is the distance from the center of the tube to the center of the torus, and

a is the radius of the tube.

An implicit equation in Cartesian coordinates for a torus radially symmetric about the z-axis is

$$\left(\sqrt{x^2 + y^2} - \rho\right)^2 + z^2 = a^2. \tag{2.4}$$

Exercise 15. Check (2.4) for (x, y, z) given by (2.3).

Fix now an initial point (θ, φ) on the torus and consider its motion, depending on time t, given by the rule

$$\theta_1 = \alpha + \theta t \quad \text{and} \quad \varphi_1 = \beta + \varphi t, \tag{2.5}$$

where (α, β) is an initial point. We allow coordinates θ_1 and φ_1 to vary arbitrarily (may be larger then 2π and a negative number), however we do not forget that for any integer m and n a point with coordinates $(\theta_1 + 2\pi m, \varphi_1 + 2\pi n)$ coincides with a point with coordinates (θ_1, φ_1). On the plane $O\theta_1\varphi_1$ the written equations sets a straight line $\varphi_1 = \frac{\varphi}{\theta}\theta_1 + \gamma$.

According to the fact that θ_1 and φ_1 change only from 0 to 2π, moving on this straight line, we reach one of the boundary edges of the square, i.e., torus $S^1 \times S^1 = [0, 2\pi) \times [0, 2\pi)$. This point "jumps" in the corresponding point of the opposite edge of the square, then continues to move on the square in the same direction, as earlier, up to the following jumping. Thus, the trajectory on the square consists of intervals, parallel to each other. After a square gluing together in a torus these intervals (pieces) stick together to the continuous curve at a torus - it winds a torus with "angular" speeds θ and φ. This curve is called a winding trajectory with frequencies (θ, φ).

Let us find out when a torus winding trajectory with frequencies (θ, φ) is periodic.

If $\theta = 0$, then a trajectory is periodic with period $T_2 = \frac{2\pi}{\varphi}$. If $\varphi = 0$ then a periodic movement also occurs. Assume now that $\theta \neq 0$ and $\varphi \neq 0$. Then after some time multiple to $T_1 = \frac{2\pi}{\theta}$, the moving point appears on the same median of the square, and after some time multiple to $T_2 = \frac{2\pi}{\varphi}$, the point appears on the same parallel. The motion is periodic if and only if after some time T, it appears in the initial place. For this it is necessary that the time T is a multiple of both T_1 and T_2, i.e., $T = mT_1$ and $T = nT_2$ with $m, n \in \mathbb{Z}$. Thus we have the following.

Theorem 2.4. *The dynamical system on the torus given by (2.5) is periodic if and only if $\frac{\varphi}{\theta}$ is a rational number.*

Proof. Follows from the above mentioned property that there are natural numbers m, n such that $mT_1 = nT_2$, i.e.,

$$m\frac{2\pi}{\theta} = n\frac{2\pi}{\varphi} \Leftrightarrow \frac{\varphi}{\theta} = \frac{n}{m} \in \mathbb{Q}.$$

$\square$

The following theorem is about the case when $\omega = \frac{\varphi}{\theta}$ is an irrational number.

Theorem 2.5. *If $\omega = \frac{\varphi}{\theta}$ is an irrational number then any trajectory of the dynamical system on the torus, given by (2.5), is dense on the tours, i.e. the trajectory passes any small domain on the torus.*

Proof. Let U be a domain on the torus, consider a meridian S of the torus which has an arc Δ as intersection, i.e. $\Delta = U \cap S$ (see Fig. 2.9). Consider now the part $P_0, P_1, \ldots$ of winding trajectory which lie on the meridian S. A point moving along a trajectory, comes back to the meridian S through the time terms equal to $T_1 = \frac{2\pi}{\theta}$. For each such time term the longitude of a point of φ_1 changes at the same value $\hat{\varphi} = \varphi T_1 = 2\pi\frac{\varphi}{\theta} = \omega$. It means that the point P_n of the intersection of the trajectory with the meridian S is obtained from the previous such point P_{n-1} by a rotation on angle $\hat{\varphi} = 2\pi\omega$ along the meridian S. In case of irrationality of ω from Theorem 2.1, it follows that at least one of the points P_n will pass Δ. Thus, the trajectory will pass the domain U. $\square$

Let us give some useful remarks:

Remark 2.3.

1. Note that the sequence $\{P_n\}$ of points on a meridian S is everywhere dense and is uniformly distributed (i.e. the proportion of

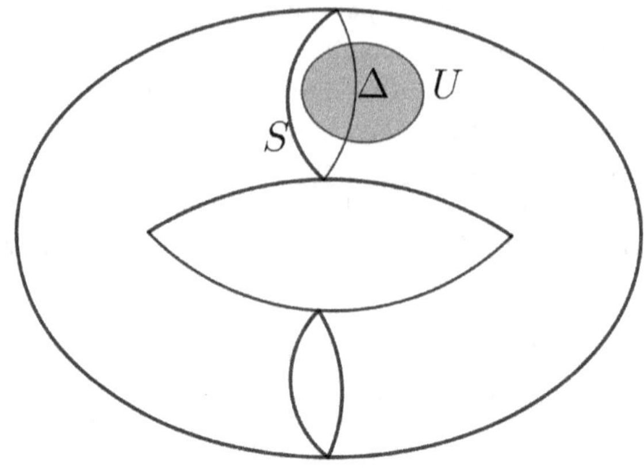

Fig. 2.9   The meridian S of the tours and the arc $\Delta = U \cap S$.

terms of $\{P_n\}$, falling in $\Delta \subset S$ is proportional to the length of Δ) and a similar assertion is true for any parallel at the torus. A torus winding trajectory with irrational ω fills a torus uniformly, i.e. the point moving uniformly along this trajectory spends a time, in given domain U of the torus, proportional to the area of U.

2. Above we talked only of winding trajectories, but having fixed winding frequencies θ and φ, we can carry out a trajectory of this winding through each point of a torus and consider all these trajectories simultaneously. They form a so-called flow of trajectories, or a torus winding. If $\omega = \frac{\varphi}{\theta}$ is a rational number, then the winding is called a rational winding , otherwise it is called an irrational winding of a torus. All trajectories of a rational winding of a torus are periodic with the same period, and any trajectory of an irrational winding everywhere densely fills the torus.

3. As mentioned above, it is possible to present a torus as a set of points (θ, φ), where θ and φ, independent of each other, vary arbitrarily in $[0, 2\pi]$, and the torus $\{(\theta, \varphi)\}$ is a configuration space of these points. Thus, we have that "the torus is a direct product of circles" and write down it in the form $\mathbb{T}^2 = S^1 \times S^1$. Similarly, if $n > 2$ circles are given on which points P_1, P_2, $\ldots, P_n$ uniformly move, then their configuration space will be the surface of an n-dimensional torus $\mathbb{T}^n = S^1 \times S^1 \times \cdots \times S^1$ (n times), located in $2n$-dimensional space $\mathbb{R}^{2n}$. A point $\theta = (\theta_1, \theta_2, \ldots, \theta_n)$

moves on this torus, where $\theta_i \in [0, 2\pi]$, $i = 1, 2, \ldots, n$. If ω_1, ω_2, $\ldots$, ω_n are the angular speeds of the points P_1, P_2, $\ldots$, P_n on the corresponding circles, i.e.,

$$\theta_1 = \omega_1 t + \alpha_1, \ldots, \theta_n = \omega_n t + \alpha_n,$$

where $\alpha_1, \alpha_2, \ldots, \alpha_n$ are the initial points on the corresponding circles. Thus the point $\theta(t) = (\omega_1 t + \alpha_1, \ldots, \omega_n t + \alpha_n)$ describes the winding trajectory on the torus $\mathbb{T}^n$.

The following theorem is a generalization of Theorem 2.4 and 2.5 from a two-dimensional torus to the n-dimensional one.

Theorem 2.6. *The winding trajectory on the torus $\mathbb{T}^n$ is*

  a. *periodic if $k_1\omega_1 + \cdots + k_n\omega_n = 0$ for some integers $k_1, \ldots, k_n$.*
  b. *everywhere dense if $k_1\omega_1 + \cdots + k_n\omega_n \neq 0$ for any integers $k_1, \ldots, k_n$, with $k_1^2 + k_2^2 + \cdots + k_n^2 \neq 0$.*

Problem 2.8.

  a) *Prove that there is a natural number n such that numbers 2^n and 3^n start with digit 7 at the same time.*
  b) *Prove that the numbers 2^n and 5^n may only start with 3 at the same n.*

Proof. a) We have to show (see solution of Problem 2.4) that there is n such that

$$7 \cdot 10^k \leq 2^n < 8 \cdot 10^k \quad \text{and} \quad 7 \cdot 10^m \leq 3^n < 8 \cdot 10^m.$$

This system of inequalities can be written as

$$k \leq \log 2 - \log 7 < k + \log(8/7) \quad \text{and} \quad m \leq \log 3 - \log 7 < m + \log(8/7).$$

Consider the circle S^1 of length 1, i.e., $S^1 = [0, 1)$, let

$$P_n = P_{n-1} + \log 2 (\mathrm{mod}\, 1) \quad (\text{resp. } Q_n = Q_{n-1} + \log 3 (\mathrm{mod}\, 1)), \ n \geq 0,$$

be a billiard trajectory with the rotation angle $\omega_1 = \log 2$ (resp. $\omega_2 = \log 3$) with $P_0 = Q_0$. Then we can take $\mathbb{T}^2 = S^1 \times S^1$ as a configuration space of points $V_n = (P_n, Q_n)$. Here the angular speed corresponding to the coordinates of the point V_n are $\log 2$ and $\log 3$ respectively. Note that $\{V_n\}$ generates a discrete-time dynamical system. For dynamical system we are not able to apply the torus winding Theorem 2.6 which holds for the continuous-time dynamical system. To make our dynamical system a continuous-time dependent, we additionally take third circle and consider on it a continuously moving point Y_0 with a constant rotation angle $\omega_3 = 1$. The configuration space of the three trajectories is now three-dimensional torus $\mathbb{T}^3$ on which now the movement is continuous and makes the torus winding. This winding is dense in the torus $\mathbb{T}^3$ iff $\lambda = k\omega_1 + m\omega_2 + l \neq 0$ (see Theorem 2.6) for any integers k, m, l. If this λ may be zero for some k, m, l then $10^\lambda = 2^k 3^m 10^l = 2^{k+l} 3^m 5^l = 1$. Since $2, 3, 5$ are coprime, from the last equality we get $k = m = l = 0$, i.e. there are no integers making λ equal to zero. Thus, the winding is dense in $\mathbb{T}^3$, and consequently, the intersection of the winding trajectory with $\mathbb{T}^2$ is also dense. This completes the proof of a).

b) Since $2^5 = 32$ and $5^5 = 3125$ the numbers 2^n and 5^n may start with digit 3 at the same n. But these numbers do not start with the same digit $a \neq 3$. Indeed[3], assume there is $a \in \{1, 2, \ldots, 9\}$ such that $2^n = a\ldots$, $5^n = a \ldots$. It is easy to check that for $n = 1, 2, 3, 4$ there is no such a. In case $n \geq 5$ we represent the numbers as

$$2^n = \alpha \cdot 10^m, \quad 5^n = \beta \cdot 10^k, \tag{2.6}$$

where α and β in decimal have the following form

$$\alpha = 0, x \ldots, \quad \beta = x, \ldots \tag{2.7}$$

Since $n \geq 5$ we have $m \geq 1$ and $k \geq 1$. From (2.6) we get $10^n = \alpha\beta 10^{m+k}$, i.e., $\alpha\beta = 10^s$ with $s = n - m - k$. On the other hand from (2.7) we have $\frac{1}{10} < \alpha\beta < 10$. Consequently, we get

$$\alpha\beta = 1, \tag{2.8}$$

i.e., $s = 0$, which means $n = m + k$. Now, we consider the following cases

$a < 3$: then by (2.7) we get $\alpha\beta < 1$ contradicting (2.8).

$a > 3$: then by (2.7) we get $\alpha\beta > 1$ contradicting (2.8). $\square$

Exercise 16. a) For a given number a, prove that there is a natural number n such that numbers 2^n and 3^n start with a at the same time.

b) For a given number a, prove that 2^n, 3^n and 7^n may start with a at the same n.

Exercise 17. [30] Consider a billiard trajectory in the unit circle, where at each impact the trajectory makes angle α with the circle.

(a) Find the central angle θ from the circle's center, between each impact point and the next one, as a function of α.

---

[3] This is proof of my PhD student I.A. Sattarov.

(b) Prove that if $\theta = 2\pi p/q$ for some $p, q \in \mathbb{N}$, then every billiard orbit is q-periodic and makes p turns around the circle before repeating.

(c) Prove that if θ is not a rational multiple of π, then the orbit of every point is dense: every interval on the circle contains points of its orbit.

2.4 Problems on 2-periodic trajectories

Let D be a planar billiard table with a smooth boundary. A 2-periodic billiard trajectory (orbit) is a segment inscribed in D which is perpendicular to the boundary at both end points. This means that the segment lies on the normal at the endpoints. Therefore 2-periodic points are related to properties of the normals of the boundary.

The following problem about a family of curves having a nice property of normals:

Problem 2.9. *Find a family of smooth, finite curves (given on the plane) all normals of which have a unique intersection point.*

Solution. Assume a curve γ is the graph of a differentiable function $y = f(x)$, given on an interval $A \subset \mathbb{R}$, i.e.

$$\gamma = \{(x, f(x)) : x \in A\}.$$

Let $(a, b) \in \mathbb{R}^2$ be the point where all normals intersect. Denote $K = \{x \in A : f'(x) = 0\}$. Equation of the normal of γ at arbitrary point $(u, f(u)) \in \gamma$ has the following form

$$y = -\frac{1}{f'(u)}(x - u) + f(u), \quad \text{if } u \in A \setminus K,$$
$$x = u, \quad \text{if } u \in K. \tag{2.9}$$

Since the normals intersect at (a, b) from (2.9), for arbitrary $u \in A$ we get

$$b = -\frac{1}{f'(u)}(a - u) + f(u), \quad \text{if } u \in A \setminus K,$$
$$a = u, \quad \text{if } u \in K. \tag{2.10}$$

Therefore, the unknown function $y = f(u)$ satisfies the differential equation

$$b = -\frac{1}{y'}(a - u) + y \quad \Rightarrow \quad (b - y)dy = -(a - u)du \tag{2.11}$$

From (2.11) we get $(u - a)^2 + (y - b)^2 = C$, where $C > 0$ is an arbitrary number. Thus the curve is a circle with the center (a, b) and with an arbitrary (finite) radius.

By the solution of Problem 2.9, we obtain the following

Proposition 2.1. *If all normals of a smooth, finite curve γ intersect at a unique point, then the curve is an arc of a circle and the intersection point is the center of the circle.*

Exercise 18. Find smooth, finite curves (if any) where all tangents have a unique intersection point.

Problem 2.10.

 a) *Does there exist a domain D without a 2-periodic (resp. 3-periodic) billiard trajectory?*
 b) *Assume that D is convex. Show that there exist at least two distinct 2-periodic billiard orbits in D.*
 c) *Are there planar billiard tables containing an infinite family of 2-periodic billiard trajectories?*

Solution. a) The answer is given in [110]: In the Fig. 2.10 the first does not have 2-periodic trajectories and the second, does not have 3-periodic trajectories. In [10], it is shown that a planar domain with a smooth boundary has either a 2- or 3-periodic billiard trajectory.

Fig. 2.10   The billiard tables without 2-periodic (left) and 3-periodic (right) trajectories.

b) Let γ be a smooth strictly convex billiard curve. A 2-periodic billiard trajectory is a chord of γ which is perpendicular to γ at both end points. Such chords are called diameters. One such diameter is easy to find: consider the longest chord of γ. Since billiard trajectories are extrema of the perimeter length function (see Chapter 4, subsection 4.2.2), the maximal chord is a 2-periodic trajectory. Now, we shall show that there is another 2-periodic trajectory. The example of an ellipse suggests that, along with

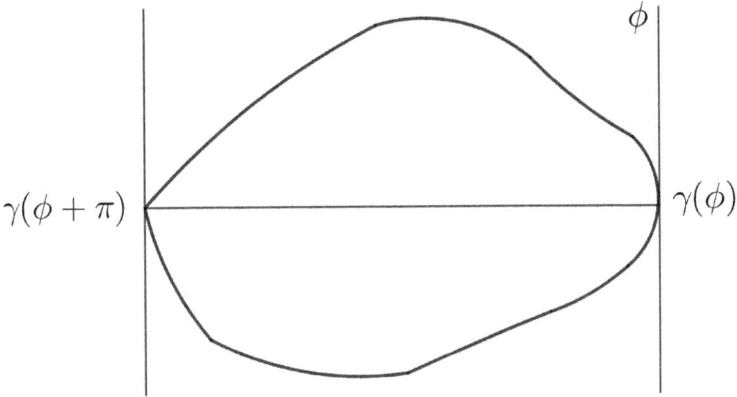

Fig. 2.11   The width of the billiard table.

the major axis, there is a second diameter, the minor axis. To construct this second diameter for an arbitrary γ, consider two parallel support lines to γ having direction ϕ (Fig. 2.11). Let $w(\phi)$ be the distance between these lines, that is, the width of ϕ in the direction ϕ. Then $w(\phi)$ is a smooth and even function on the circle. Its maximum corresponds to the longest chord of γ, and its minimum to another diameter, the desired second 2-periodic billiard trajectory.

c) Very simple example is a disc D in the plane which contains a one parameter family of 2-periodic billiard trajectories making a complete turn inside D, i.e., these trajectories are the diameters of D. Now using Proposition 2.1 one can construct billiard tables (different from a disc) with infinitely many 2-periodic points: take concentric discs with distinct radiuses and combine their arcs to construct the required tables (see Fig. 2.12).

Exercise 19. a) Does there exist a domain D without a 4-periodic billiard trajectory?

b) Construct planar billiard tables (different from above-mentioned tables) containing an infinite family of 2-periodic billiard trajectories.

c) Construct planar billiard tables containing infinite family of 3-periodic trajectories.

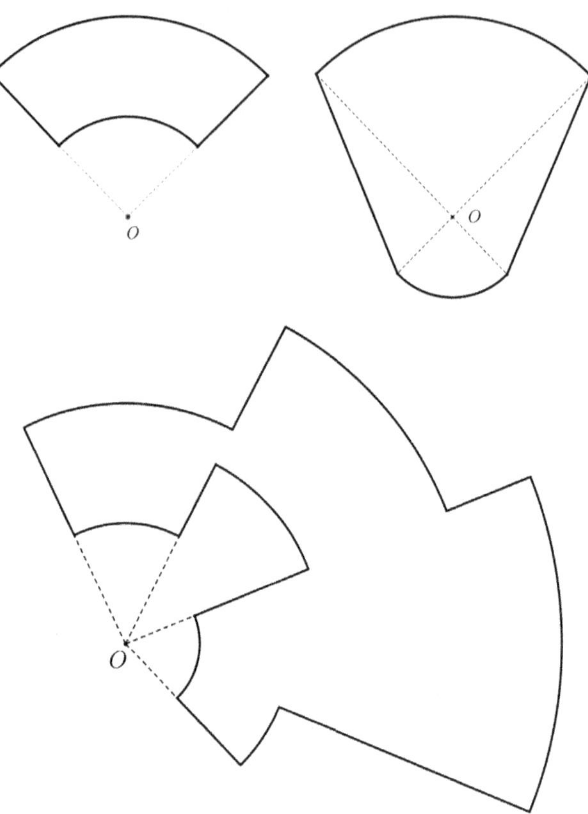

Fig. 2.12   The billiard tables with infinitely many 2-periodic trajectories. The boundary of the tables are combined by arcs of concentric circles.

2.5   The number π from a billiard point of view

2.5.1   *The procedure*

Many ways are known to calculate π with a good precision; some of them are known from ancient times, some are recent. History of π is given in [9]: the methods use various elegant ideas such as

> *geometric*: inscribing and circumscribing regular polygons around a circle gives, in particular, the ancient values $3\frac{1}{7}$ and $3\frac{10}{71}$ for π;

number theory: continued fractions allow us to find the regular fraction $\frac{355}{113}$ as the simplest approximation for π accurate to the one millionth place);

analytical: that uses series, integrals, and infinite products;

experimental: the Buffon's method[4], for finding π, he suggested dropping a needle of length $L = D/2$ at random on a grid of parallel lines of spacing D. One drops the needle N times and counts the number of intersections, R, with the grid lines.[5] The frequency of intersection with a line is R/N; on the other hand, one can show that the probability for the needle to intersect a grid line is $1/\pi$.[6] Taking the frequency and the probability the same, we obtain that π approximately equals N/R, the ratio of the number of drops to the number of intersections with the grid.

others: e.g., the Monte Carlo Method which requires modern electronic devices: powerful calculators and computers.

In this section following [42] we give a new procedure for calculating π using mathematical billiards. This procedure is similar to the Buffon's method (i.e. also experimental), but does not require use of any computer devices. In contrast to the Buffon's method, this procedure is entirely deterministic. This is related to the dynamical system consisting of two billiard balls and an absolutely elastic obstacle (a wall). One has to count the number of collisions in that system, and then write down this number on a sheet of paper. The integer that is going to be written will be

$$31415926535897932384626433832795028841971693993751 0 \ldots$$

It consists of the first N digits of $\pi = 3.14159265\ldots$, where N represents the number of decimal digits of π one wants to know.

The procedure is given as follows. Consider two point-like balls with masses m and M, $M \geq m$. The balls will move along the positive x-axis and collide with each other at every encounter, and the small ball, m, will reflect off a vertical wall located at point $x = 0$.

We assume that each collision in the system is absolutely elastic, meaning that a collision between the balls satisfies two mechanical laws: the law

---

[4] http://www.cut-the-knot.com/fta/Buffon/buffon9.htm

[5] Since the needle is shorter than the distance between two consecutive grid lines, it intersects each time either exactly one line or none of the lines.

[6] For an arbitrary needle of length L, the probability equals $\frac{2L}{\pi D}$.

of conservation of momentum, and the law of conservation of kinetic energy. Moreover, the small ball reflects from the wall by changing its velocity vector to the opposite vector. In other words, the wall can be thought of as a non-moving billiard ball of infinite mass (Fig. 2.13).

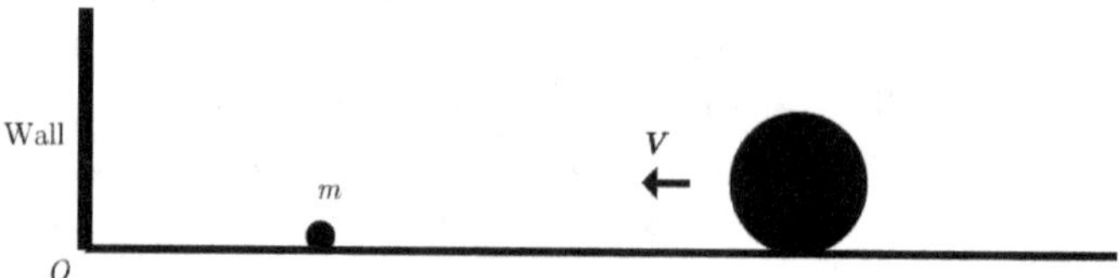

Fig. 2.13   The system with to balls and the ball M pushing towards the ball m.

The procedure consists the following steps:

1) Let N be a fixed positive number. Consider two billiard balls with the ratio of their masses $\frac{M}{m} = 100^N$.
2) Put the small ball, m, between the wall at the origin and the big ball, M.
3) Push the big ball towards the small ball.
4) Calculate the total number of hits in the system: the number of collisions between the balls plus the number of reflections of the small ball from the wall.
5) Write down the number Π of hits obtained from the previous step (i.e. 4).

For different values of N this Procedure gives us different values for the number Π. Thus $\Pi = \Pi(N)$ is a function of the exponent N of the number 100^N.

The simplest case is $N = 0$, which corresponds to the equality of the masses: $M = m$. The laws of conservation yield the following description of the systems behavior:

1) M hits m and stops; m begins to move to the left;
2) m hits the wall and bounces back;
3) m hits M and stops; M begins to move to the right and goes to infinity. Thus, the total number of hits in the system with $M = m$ is 3: two collisions and one reflection, i.e. $\Pi(0) = 3$. (See Fig. 2.14 to count geometrical hits.) Note that 3 is the first digit of π. In what follows, the number of hits, Π, is 31 (two first digits of) for $M = 100m$. (See Fig. 2.14 for a geometric representation.)

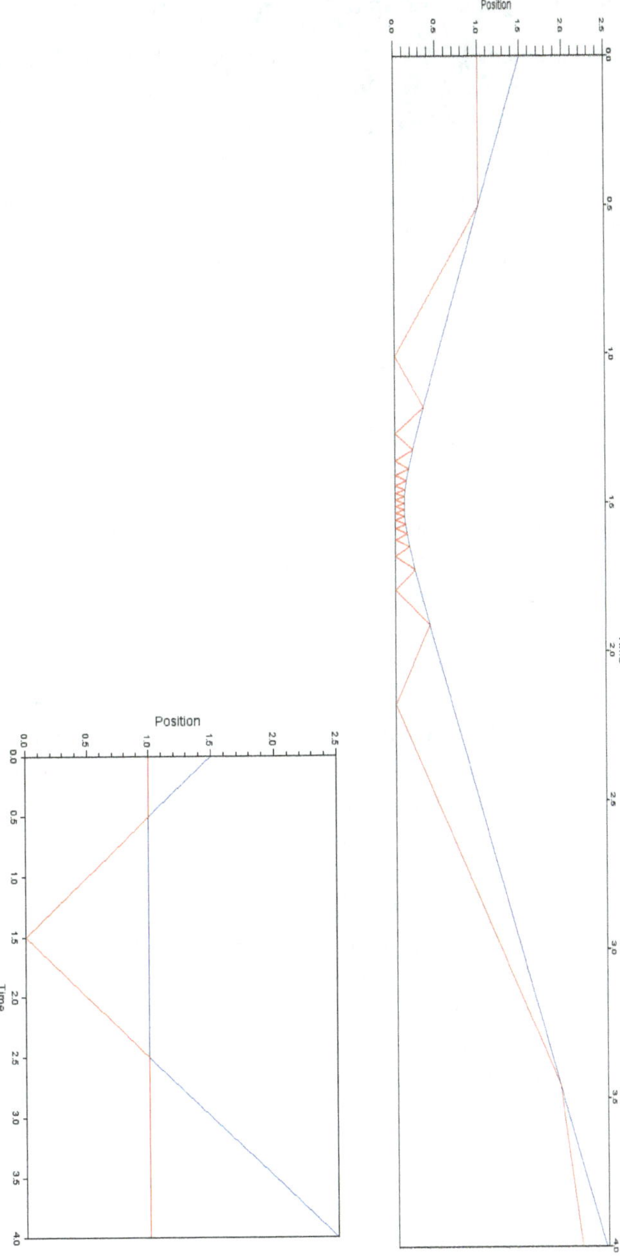

Fig. 2.14 Small ball at $x = 1$ and ball M at $x = 1.5$, with initial velocity 1. The time axis is vertical. $N = 0$ (left), $N = 1$ (right). Source: Internet.

2.5.2   *The configuration space of the system*

At the initial moment of time, $t = 0$, we assume that the balls m and M are located at points x_0 and y_0 on the horizontal line ℓ which is the positive part of the x-axis. Depending on time $t \geq 0$, the coordinates of the moving balls change. Denote their states by $x = x(t)$ and $y = y(t)$, respectively.

We have $x(0) = x_0, y(0) = y_0$. By the assumptions of the procedure, at each moment t, the small ball is situated between the wall and the big ball M. This means that

$$0 \leq x(t) \leq y(t), \quad \text{for all } t \geq 0.$$

At a moment $t = t_r$ of reflection of the small ball m from the wall, this ball is always situated at the origin: $x(t_r) = 0$.

Thus, we have the following set as where all states of the system belong

$$\mathbb{K} = \{(x, y) \in \mathbb{R}^2 : 0 \leq x \leq y\}.$$

Each point of $\mathbb{K}$ is called the configuration point and the set of all possible configuration points is called the configuration space. Hence, the configuration space for the dynamical system of the two balls in question is the 45^0 angle with its interior formed by the positive y-axis and the angle bisector of the first quadrant of the xy-plane, i.e. the ray $\{(x, y) : \quad x = y, x \geq 0\}$.

Therefore, along with the initial physical system of two balls on the semi-line, we can consider another mathematical system, i.e., the mathematical model of the initial system. Both of the systems are dynamical systems; they are different formal descriptions of the same phenomena.

Now we study the behavior of the configuration point $P(t) = (x(t), y(t))$. We should make following steps:

Step 0: **Before the first collision.** At the initial moment $t = 0$, the configuration point P is located on the plane at the geometric point $P_0 = (x_0, y_0)$. When the time changes the point $P = P(t)$ starts moving. The small ball is fixed before the first collision, therefore the x-coordinate of the moving point P does not change. On the other hand, the big ball M moves towards the small ball m, and hence its coordinate y decreases, remaining, however, bigger than x_0 during the entire time period before the first collision with m. Consequently, the configuration point P moves directly down toward the x-axis (parallel to the y-axis) until the first collision.

Step 1: **First collision.** The big ball collides with the small one; at this moment, t_1, i.e., $y(t_1) = x_0$. Then the balls bounce off each other instantaneously, and the next step begins.

Step 2: **Between the first collision and the first reflection.** At the moment t_1, both balls begin moving along the horizontal line ℓ. If the small ball moves with some velocity u and the big ball with velocity v, then the laws of conservation of momentum and energy hold:

$$mu + Mv = MV,$$
$$mu^2 + Mv^2 = MV^2, \tag{2.12}$$

here V is the initial velocity of the ball M. We are interested to describe the behavior of the configuration point after the first collision. Using the system of equations (2.12), we conclude that, after the first collision, the ball m will move very fast towards the wall (since the big ball gives it a big momentum) and the ball M also continues to move, a little bit slower than before, towards the wall. Both coordinates $x(t)$ and $y(t)$ are decreasing on the time interval after the first collision but before the reflection of the ball m from the wall. Therefore the configuration point $P(t) = (x(t), y(t))$ moves along a straight line segment inside the angle $\mathbb{K} = AOB$, where O is the origin, OA is the positive y-axis, and OB is the ray $y = x$ outgoing from the origin in the first quadrant. Point $P(t)$ travels from the side OB to the side OA, approaching the origin O.

Step 3: **Reflection from the wall.** The small ball m moves faster than the big ball after the first collision, i.e. $u > v$. Because, first note that $v < V$, since the big ball gives some momentum to the small ball and accelerates it, so its new velocity v becomes smaller. Multiplying the first equation of the system (2.12) by V and subtracting the second equation yields

$$Mv(V - v) = mu(u - V).$$

Since $Mv > 0$, $V - v > 0$, $mu > 0$ we get from the last equality that $u > V$. Consequently, $u > V > v$. For the convenience, we assume the velocity of a ball moving from right to left to be positive, and from left to right to be negative. To do this assumption, one can consider speeds (the absolute values of velocities) instead of the velocities.

Therefore, the ball m reaches the wall at some moment t_2 at which the ball M is still moving toward the wall. At t_2, the ball reflects off the wall, and its velocity instantaneously jumps from u to $-u$. The momentum of the small ball becomes $-mu$, but its energy remains the same: $m(-u)^2/2 = mu^2/2$.

The angle of incidence and the angle of reflection formed by the segments of the trajectory with the x-axis is equal.

Step 4: **After the first reflection off the wall.** The point $P(t)$ moves after the first reflection from the y-axis along a straight line segment towards the line $y = x$ approaching the origin O: its x-coordinate increases and the y-coordinate decreases. This corresponds to the balls approaching each other when the ball m bounces off the wall.

Since the wall is not considered part of the ball system, the momentum of the system after the wall reflection has been changed from $mu + Mv = MV$ to $-mu + Mv$, but the energy remains the same, $MV^2/2$. The momentum does not change between two successive reflections off the wall (see Fig. 2.15).

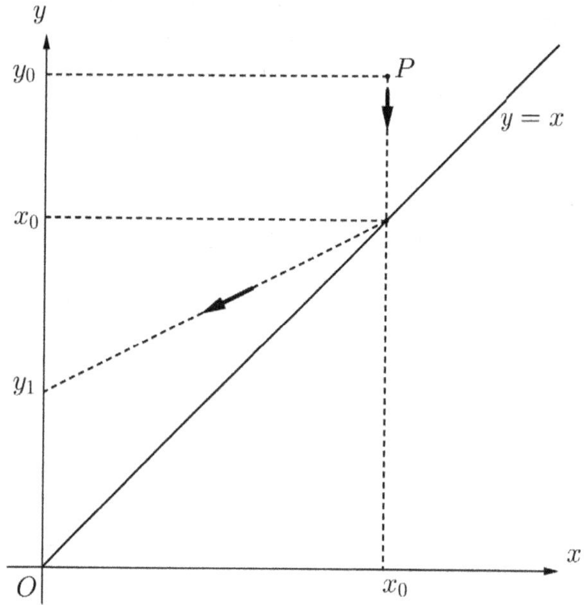

Fig. 2.15　Initial position P and the direction after the first collision.

Step 5: **The second collision of the balls.** At the moment when P reaches the side $y = x$ of the configuration angle AOB, the second collision of the balls occurs. Assume it happen, at a point $x_1 > 0$ on line ℓ. The balls change their velocities after the collision. Denoting the new velocities by u_1 and v_1, we get

$$mu_1 + Mv_1 = -mu + Mv,$$
$$mu_1^2 + Mv_1^2 = Mv^2. \tag{2.13}$$

Similar to the given above case, we have $|u_1| > |v| > |v_1|$.

Steps 6, 7,...: Starting from Step 5, the initial situation repeats: The small ball, m, moves with velocity u1 towards the wall, while the big ball, M, reduces its velocity from v to v_1 but continues to move towards the wall; The small ball reflects from the wall, changes its velocity, u_1, to the opposite $-u_1$; The small ball meets the big ball and collides with it.

Iterating the process one can see that the behavior of the system of the balls is reflected in the motion of the configuration point P as shown in Fig. 2.16.

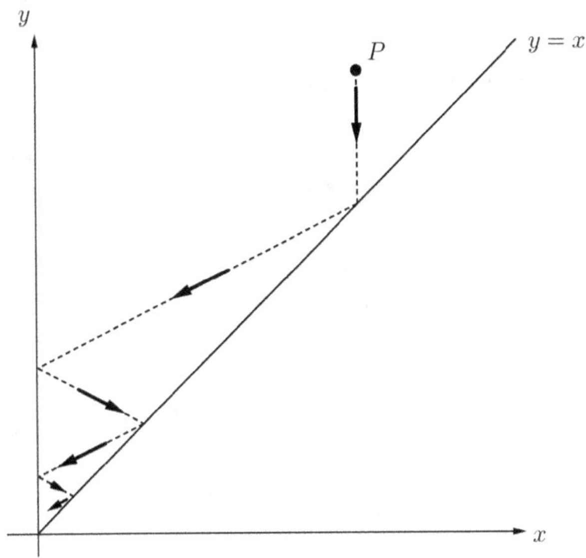

Fig. 2.16   Initial position P and the configuration path.

2.5.3 *Behavior of the dynamical system*

In general, P may have one of the following kind behaviors:

(i) P approaches the vertex O of the configuration angle AOB forever;

(ii) P approaches O for a finite period of time, then moves away from O and reflects off the sides of the angle AOB infinitely many times;

(iii) P makes only finitely many reflections off the angles sides, and, from some moment T_0 moves freely and rectilinearly.

Note that the Cases (i) and (ii) correspond to *infinitely* many hits in the system, i.e., collisions between the balls and reflections from the wall, and the Case (iii) to *finitely* many collisions and reflections. The following theorem states that only the Case (iii) can occur, where the total number of hits in the system is $\Pi = 314159265\ldots$ depending on N.

Theorem 2.7. [42] *The number of hits, $\Pi = \Pi(N)$, in the system described in the Procedure is always finite and equal to a number with $N + 1$ digits,*

$$\Pi(N) = 3141592653589793238462643383279502884197169399375 10 \ldots ,$$

whose first N digits coincide with the first N decimal digits of the number π (starting with 3).

Proof. Following page 6 of [110] we give a shortened proof. For a detailed proof see [42]. Given an initial point, the configuration trajectory enters the angle in the direction, parallel to the vertical side. In this case, the number of reflections is given by (see formula (8.1) in [42])

$$\Pi(N) = \left\lceil \frac{\pi}{\arctan(10^{-N})} \right\rceil - 1,$$

where $\lceil x \rceil$ is the ceiling function which maps x to the least integer greater than or equal to x.

Denote $x = 10^{-N}$. This x is a very small number, and one expects $\arctan x$ to be very close to x. Namely, using the Taylor expansion for the function $\arctan x$ we get

$$0 < \frac{1}{\arctan x} - \frac{1}{x} < x, \quad \text{for} \quad x > 0. \tag{2.14}$$

The first N digits of the number

$$\left\lceil \frac{\pi}{x} \right\rceil - 1 = \left\lceil 10^N \pi \right\rceil - 1$$

coincide with the first $k + 1$ decimal digits of π. Consequently, we should show that

$$\left\lceil \frac{\pi}{x} \right\rceil = \left\lceil \frac{\pi}{\arctan x} \right\rceil. \tag{2.15}$$

By (2.14) we have

$$\left\lceil \frac{\pi}{x} \right\rceil \leq \left\lceil \frac{\pi}{\arctan x} \right\rceil \leq \left\lceil \frac{\pi}{x} + \pi x \right\rceil. \tag{2.16}$$

Note that the number πx has $N - 1$ zeros after the decimal dot, i.e.

$$\pi x = 0.0\dots031415\dots$$

Therefore, the left- and the right-hand sides in (2.16) can differ only if there is a string of $N - 1$ nines following the first $N + 1$ digits in the decimal expansion of π. It is not clear whether such a string ever occurs, but this is extremely unlikely for large values of N. If one does not have such a string, then both inequalities in (2.16) are equalities, (2.15) holds, and the proof is completed. $\qquad\square$

2.5.4 *The reduction to the billiard system*

The above considered dynamical system can be reduced to the billiard system in an angle. Now we are going to explain this reduction. The motion of the configuration point P in the 45^0 angle AOB will be reduced to a billiard problem in some other angle α.

We note that a point in a billiard system behaves as a ray of light in a room (domain) with mirror walls (boundary). In order to carry out the reduction to a billiard problem, let us make a special linear transformation, T, of the xy-plane.

$$T : \quad \begin{cases} X = \sqrt{m}\cdot x \\ Y = \sqrt{M}\cdot y. \end{cases} \tag{2.17}$$

This transformation was considered first in [103]. The linear transformation T maps the 45^0 angle AOB into the angle $\alpha = A'O'B'$ satisfying

$$\tan\alpha = \frac{X}{Y} = \sqrt{\frac{m}{M}},$$

since $y = x$ for the points on the oblique side of the angle AOB.

The broken line corresponding to the trajectory of the configuration point $P(x,y)$ inside the angle AOB will be mapped into a broken line corresponding to the trajectory of the point

$$P' = P'(X,Y) = P(\sqrt{m}\cdot x, \sqrt{M}\cdot y)$$

inside the angle $\alpha = A'O'B'$.

Proposition 2.2. *The behavior of the new configuration point P' inside angle α obeys the billiard law.*

Proof. If point $P(x, y)$ has the velocity vector $\overrightarrow{\mathbf{w}} = (u, v) = (\dot{x}(t), \dot{y}(t))$ at moment t, then point $P'(t)$, at moment t, has the velocity vector

$$\overrightarrow{\mathbf{v}} = (\sqrt{m}u, \sqrt{M}v) = (\sqrt{m}\dot{x}(t), \sqrt{M}\dot{y}(t)).$$

Thus the linear transformation T of the configuration space $\mathbb{K}$ induces the same linear transformation in the velocity space $\{(\dot{x}, \dot{y})\}$.

Consider the following two cases:

Case 1: *Reflection from the y-axis.* When the small ball reflects from the wall, its velocity u changes to $-u$. Then vector $\overrightarrow{\mathbf{v}}$ converts into vector $\overrightarrow{\mathbf{v}'} = (\sqrt{m}(-u), \sqrt{M}v)$ which means that the y-axis is the bisector of the angle made by $\overrightarrow{\mathbf{v}}$ and $\overrightarrow{\mathbf{v}'}$, i.e. the billiard reflection law holds (see Fig. 2.16).

Case 2: *Reflection from the side $Y = \sqrt{M/m}X$.*

This reflection corresponds to the ball collision. Consider an interval of time in which only this collision occurs, i.e., the interval between two successive reflections of the ball m from the wall.

The system of moving balls has unchanging momentum during this interval of time, and the collision of the balls does not change it. The energy is always constant during the whole process. Denote the momentum by C_1 and twice the energy by C_2. Assume the small ball has velocity u and the big ball has velocity v. The system (2.12) can be written as follows:

$$mu + Mv = C_1,$$
$$mu^2 + Mv^2 = C_2. \tag{2.18}$$

Denoting $\overrightarrow{\mathbf{m}} = (\sqrt{m}, \sqrt{M})$ and $\overrightarrow{\mathbf{v}} = (\sqrt{m}u, \sqrt{M}v)$ the system (2.18) can be rewritten as

$$\overrightarrow{\mathbf{m}} \circ \overrightarrow{\mathbf{v}} = C_1,$$
$$|\overrightarrow{\mathbf{v}}| = C_2, \tag{2.19}$$

where $\circ$ is the dot product in the xy-plane, and $|\cdot|$ is the Euclidean metric on this plane.[7] Using

$$\overrightarrow{\mathbf{m}} \circ \overrightarrow{\mathbf{v}} = |\overrightarrow{\mathbf{m}}|\,|\overrightarrow{\mathbf{v}}| \cos(\varphi) = C_1, \quad |\overrightarrow{\mathbf{v}}| = C_2, \quad |\overrightarrow{\mathbf{m}}| = \sqrt{m + M}$$

and denoting $\varphi = \overparen{\overrightarrow{\mathbf{m}}, \overrightarrow{\mathbf{v}}}$ we get

$$\cos\varphi = \frac{C_1}{C_2\sqrt{m + M}} = C_3.$$

---

[7]The dot product of two vectors $a = (a_1, \ldots, a_n)$ and $b = (b_1, \ldots, b_n)$ is defined as:

$$a \circ b = \sum_{i=1}^{n} a_i b_i.$$

The Euclidean metric is $|a| = \sqrt{a \circ a}$.

After reflection, the point P' moves with a new velocity, $\overrightarrow{\mathbf{v}}'$ satisfying the same system (2.19). Therefore, the same reasoning for the angle $\psi = \widehat{\overrightarrow{\mathbf{m}}, \overrightarrow{\mathbf{v}}'}$ of P' points reflection from the α's side $Y = \sqrt{M/m}\,X$ show that $\cos\psi = C_3$ (see Fig. 2.17). Consequently, $\psi = \varphi$, and the billiard law is proven for

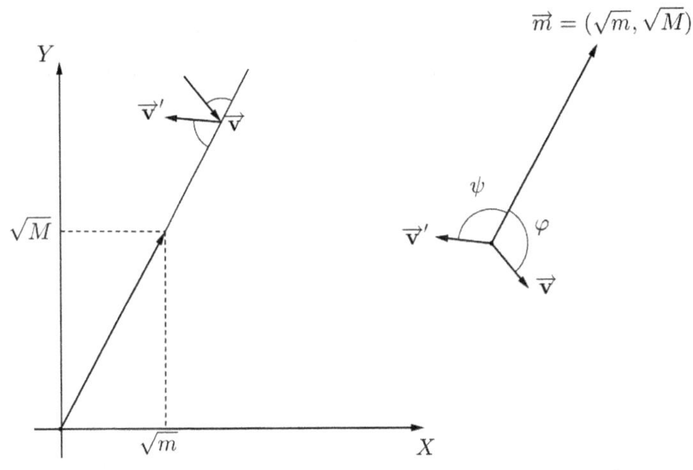

Fig. 2.17 Reflection from the side $Y = \sqrt{M/m}\,X$.

this reflection too. The reduction to the billiard system in the angle α is completed. $\qquad\square$

Consider two point-like balls with masses m and M, $M \geq m$. The ball M pushed toward the ball m and they will move along a circle S^1 of length L and collide with each other at every encounter, and the small ball, m, will reflect off a vertical wall located at a fixed point x_0 of the circle.

As before, we assume that each collision in the system is absolutely elastic, meaning that a collision between the balls satisfies two mechanical laws: the law of conservation of momentum, and the law of conservation of kinetic energy. Moreover, the small ball reflects from the wall by changing its velocity vector to the opposite vector. Thus a new system defined by the procedure consists of the following steps:

i) Consider a circle S^1 of length L, fix a point of it and put a "wall", and put on S^1 two billiard balls with the ratio of their masses $\frac{M}{m} = 100^k$, for some $k = 0, 1, 2, \ldots.$

ii) Push the big ball towards the small ball.

Exercise 20. Using laws of circular motion[8]

a) calculate the total number $\diamond(k)$ of hits in the system i), ii), which is the number of collisions between the balls plus the number of reflections of the small ball from the wall.

b) Is $\diamond(0)$ a finite number?

Bibliographical notes. There are several literature mentioning the simple applications of the mathematical billiard to elementary mathematics (see for example [24], [37]-[41], [48]).

This chapter is based on [12], [30], [37], [38], [42], [86], [91], [92], [108], [110] and many internet sources.

---

[8]https://en.wikipedia.org/wiki/Circular_motion

Chapter 3

Billiards and geometry

In this chapter we give an introductional part of the theory of mathematical billiards related to geometry of the billiard tables. We present results on the behavior of a billiard trajectory on a planar table, having one of the following forms: triangle, ellipse, rectangle, polygon and some general convex domains.

3.1 Configuration space

A configuration space of a system describes the assignments of a collection of points to positions (states, phases) of the system. This is a geometric representation of the set of all possible positions of the system.

Let us give some examples:

Example 3.1.

1. Consider a point x moving in segment $[0, 1]$, then its configuration point $x(t)$ at moment t, is given by a number $x(t) \in [0, 1]$. Thus the configuration space, of all possible positions (also called states), of the point is the segment $[0, 1]$.

2. Consider two balls (points) x and y moving independently in segment $[a, b]$, then at moment t the position of the system is given by the two-dimensional vector $(x(t), y(t))$, therefore, the configuration space is the square $[a, b] \times [a, b]$. More generally, if x is moving in $[a, b]$ but y is in $[c, d]$, then the configuration space of this system is $[a, b] \times [c, d]$.

3. Consider now two points x, y in $[0, 1]$, which move depending on each other, such that $x \leq y$. Then the corresponding configuration

61

space is the triangle (i.e. a half of square $[0,1]^2$)

$$\{(x,y) \in [0,1] \times [0,1] : x \leq y\}.$$

4. Consider again two points in $[0,1]$ which move in such a way that the distance between them is always greater than or equal to $\frac{1}{3}$. The configuration space is

$$\{(x,y) \in [0,1] \times [0,1] : |x-y| \geq \frac{1}{3}\}.$$

5. Consider now three points x, y, z moving independently in $[0,1]$, then the configuration space of this system is the cube $[0,1]^3$.
6. The configuration space of all unordered pairs of distinct points on the circle is the Möbius strip[1], see Fig. 3.1. This is a surface with only one side (when embedded in three-dimensional Euclidean space) and only one boundary. The Möbius strip has the mathematical property of being unorientable.

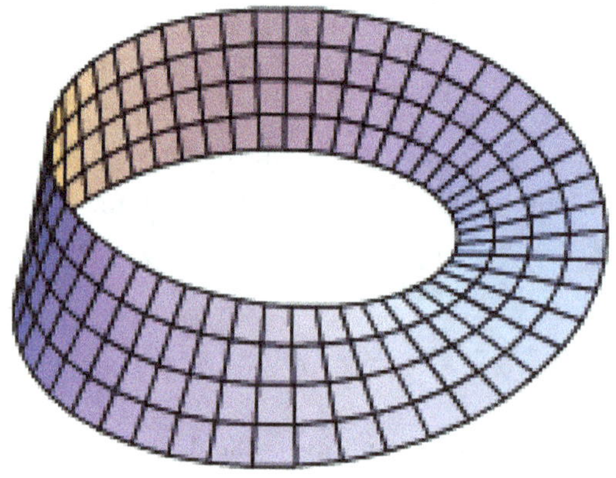

Fig. 3.1   The configuration space of two unordered distinct points on a circle.

Knowing only the configuration space of a system may be useful when solving some difficult problems. For example, let us consider the following problem of N. N. Konstantinov (see page 48 of [37]):

Problem 3.1. *From a city A to a city B there are two roads (see Fig. 3.2). Two cyclists are bound with a rope of length smaller than $2a$, and they can*

---

[1] https://en.wikipedia.org/wiki/Configuration_space_(mathematics)

pass from city A to city B on different roads, without having the rope torn off. Consider two balls, each having radius a, the first ball going, by the first road, from city A to city B, and the second ball going, by the second road, from city B to city A. Is it possible for the balls to reach their destination without a collision?

Solution. Configuration of the cyclists is given by point $x = (x_1, x_2)$ where x_1 (resp. x_2) is the position of the first (resp. second) bicycle (or ball). Thus the configuration space is given in the Ox_1x_2 plane, $\mathbb{R}^2$. If the length of the first road is equal to l_1 and the length of the second one is l_2 then the configuration space is

$$OKLM = \left\{ (x_1, x_2) \in \mathbb{R}^2 : x_1 \in [0, l_1], \quad x_2 \in [0, l_2] \right\}.$$

The cyclists will start their movement from city A, which is the point O in the configuration space and will finish their movement at city B, which is the vertex L of the configuration space. Thus, the configuration space of the cycles is the curve α from O to L. Similarly, the configuration space of the balls is the curve β connecting the vertex M with the vertex K. Since both curves are inside of $OKLM$, they have an intersection point $x = (x_1, x_2)$ in the configuration space. Note that at this point the coordinates of the bicycles and the ball coincide. Consequently, this means that the distance between balls is less than $2a$ at that point, because the cyclists keep a distance less than $2a$ without tearing off the rope. Thus, the balls will collide.

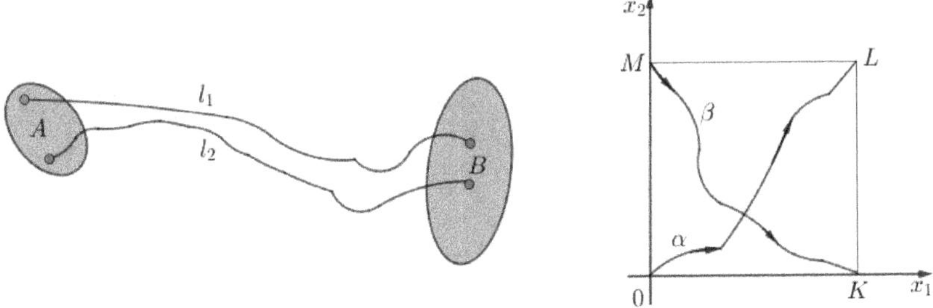

Fig. 3.2   The roads from A to B (left), and the configuration space of two moving objects (right) one from A to B and another one from B to A.

Exercise 21.

1. What is the configuration space of a point moving in a disc at a distance from the center which is greater than half of the radius of the disc?

2. What is the configuration space of two points x, y moving in segment $[a, b]$, with the condition $|x - y| = \frac{b-a}{4}$?

3. Consider two points x and y, moving on the real line $\mathbb{R} = (-\infty, +\infty)$, and assume that they satisfy $x \leq y$ and $|x - y| \leq 1$. Give the configuration space of the system.

3.2 Unfolding a billiard trajectory

A powerful tool for understanding the trajectory of a billiard is unfolding the trajectory into a straight line by creating a new copy of the billiard table each time the ball hits an edge.

The unfolding of a rectangular billiard table is introduced as follows[2]. Consider a rectangle $ABCD$ and reflect it symmetrically along the CD border, consequently producing two more vertices A' and B' and two rectangles as shown in the Figure 3.3.

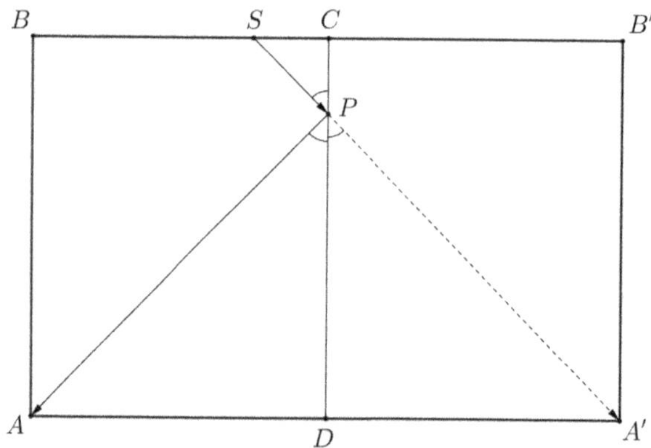

Fig. 3.3   Unfolding the billiard table.

Consider a ball at an arbitrary point S on the BC side. We wish to determine a point P on the CD side so that the ball bounces from the cushion at P and hits the corner located at A. This problem can be quite

[2]Source: see http://www.math.cornell.edu/ mec/Summer2009/Remus/lesson1.html

tedious if we do not unfold the billiard table. We unfold it as shown in the Figure 3.3. Then in order to find point P we need to intersect line SA' with CD. To describe the trajectory in the rectangle $ABCD$, we continue it in the rectangle $CDB'A'$, with the understanding that hitting corner A' would be the same as hitting A. It is easy to see that the set of rules are clearly satisfied as angles $A'PD$ and APD are the same.

Problem 3.2. *Given a square billiard table. Is it possible to hit a ball located at a given point on the lower boundary in such a way that it returns to the same point after exactly 6 bounces (as shown in Figure 3.4).*

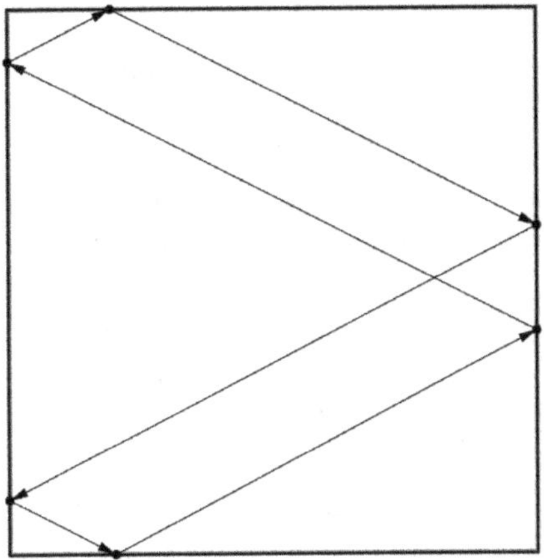

Fig. 3.4   A 6-periodic trajectory.

Solution. To answer the question, it is easier to simply unfold the billiard table, making a finite grid in the first quadrant of the plane. The two points in the Figure 3.5 are the same point, relative to the initial table.

Draw a line joining these two points and that is the trajectory of the ball in the unfolded billiard table. Obviously, it crosses exactly 5 borders, thus dividing the line segment into exactly 6 pieces. This means that the ball bounces exactly 6 times before returning to the original point.

Remark 3.1. In fact, it is possible to shoot the ball in such a way that the trajectory comes arbitrarily close to any point on the boundary, i.e. the

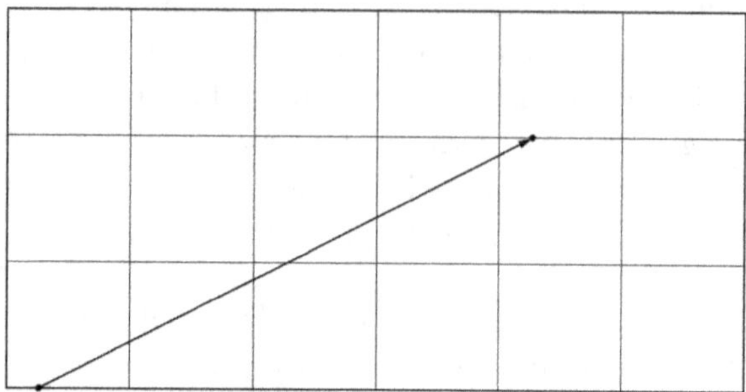

Fig. 3.5   Unfolded billiard table with a 6-periodic trajectory.

trajectory is dense. This occurs when the angle at which we shoot the ball is an irrational multiple of π.

Remark 3.2. Consider the Problem 3.2 not on a rectangular billiard table but on a circular one: Starting with a ball on the circle, we ask whether it is possible to hit it in such a way that it returns to the original position after exactly 6 bounces. As it was discussed in Chapter 2, this problem has the following answer: if the rotation angle is $\frac{p}{6}$, $p = 1, 2, 3, 4, 5$ then the trajectory will be 6-periodic (see Fig. 3.6).

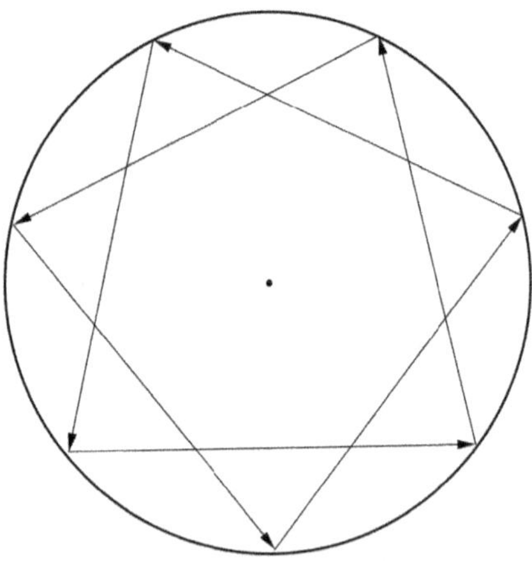

Fig. 3.6   Circular billiard table with a 6-periodic trajectory.

Remark 3.3. The unfolding construction works for any rational billiard table, that is any polygonal billiard table with angles of the form $\pi\frac{p_i}{q_i}$. For example, a billiard in a triangle with $\frac{\pi}{8}, \frac{3\pi}{8}, \frac{\pi}{2}$ angles. By unfolding this billiard trajectory, one gets linear flow in the regular octagon [116] (see Fig. 3.7).

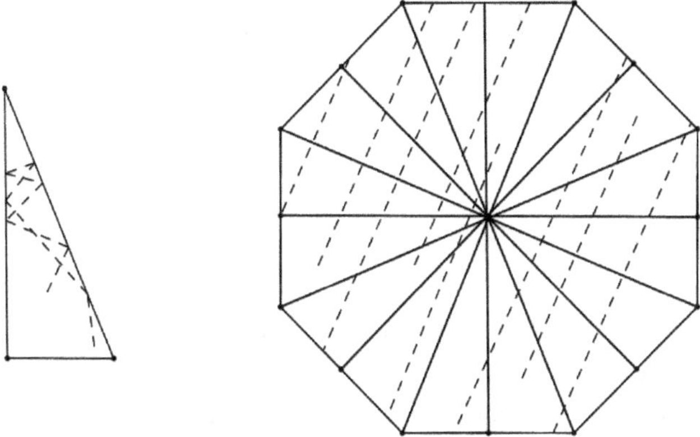

Fig. 3.7   Unfolding billiard in a triangle.

Symbolic coding of trajectories in the square. Now following [110], consider a billiard trajectory in a square having an irrational slope. Encode the trajectory with an infinite sequence (word) of two symbols, 0 and 1, according to whether the next reflection occurs in a horizontal or a vertical side respectively. Equivalently, the unfolded trajectory is a line L which meets consecutively horizontal or vertical segments of the unit grid. This sequence of zeros and ones is called the cutting sequence of the line L. Cutting sequences are Sturmian sequences.

Sturmian sequences correspond to the sequence of horizontal (symbol 0) and vertical (symbol 1) sides crossed by a line in direction θ in a square grid: $\ldots 0101101 \ldots$

Note that Sturmian sequences appear in many areas of mathematics, e.g. in Number Theory - related to the continued fractions

$$\tan\theta = \cfrac{1}{a_1 + \cfrac{1}{a_2 + \cfrac{1}{a_3 + \ldots}}};$$

in Computer Science - related to smallest possible complexity questions. The complexity of the cutting sequence of a billiard trajectory is defined as follows. Let w be an infinite sequence of some symbols (zeros and ones, in our case).

The complexity function $p(n)$ is the number of distinct segments of length n in w.

The faster $p(n)$ grows, the more complex the sequence w is. For two symbols, the fastest possible growth is $p(n) = 2^n$.

Theorem 3.1. [110] *The complexity function of the cutting sequence of a line L with an irrational slope is $p(n) = n + 1$.*

A sequence is called quasi-periodic if every one of its finite segments appears in it infinitely many times.

Theorem 3.2. *The cutting sequence w of a line L with an irrational slope is not periodic but is quasi-periodic.*

Proof. Consider a finite segment of w containing p zeros and q ones. The respective segment of L moved p units in the vertical and q units in the horizontal direction. Assume that w is periodic, and let the period contain p_0 zeroes and q_0 ones. The slope of L is the limit, as $n \to \infty$, of the slopes of its segments L_n, corresponding to the segments of w made of n periods. The slope of L_n is $(np_0)/(nq_0)$, and the limit is a rational number p_0/q_0. This contradicts our assumption that the slope of L is irrational.

If two points of the square are sufficiently close to each other, then sufficiently long segments of the cutting sequences of parallel billiard trajectories through these points coincide. Since the slope of L is irrational, it will return to any neighborhood of its points infinitely many times. Thus quasi-periodicity of w follows. $\square$

Example 3.2. [110] Consider the irrational number (known as the golden ratio) $\alpha = \frac{1}{2}(1 + \sqrt{5})$. Let L be the line through the origin with slope α. In this case, the respective cutting sequence is

$$w = ...0100101001001...$$

This sequence has the following property: w is invariant under the substitution $0 \to 01, \quad 1 \to 0$.

Symbolic coding of trajectories in the octagon. Following [106], [116] we consider a regular octagon. Glue opposite sides together. Label pairs of sides by A, B, C, D. Let f_θ^t be the linear flow in direction θ: trajectories which do not hit vertices are straight lines in direction θ.

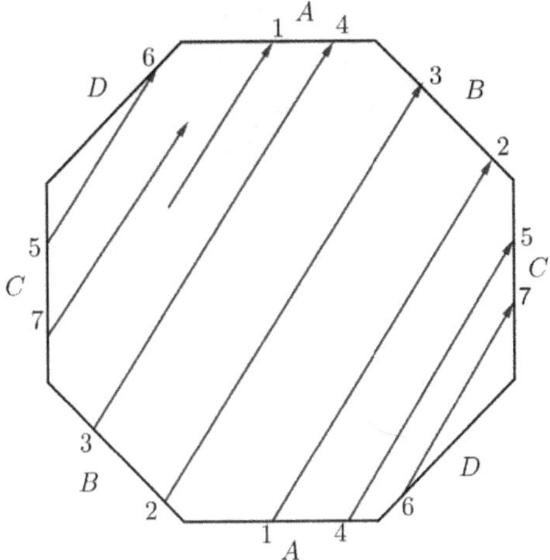

Fig. 3.8   Linear flow in octagon, sides labeled by A, B, C, D.

For given sets X and Y, denote by Y^X the set of all possible functions from X to Y.

Definition 3.1. The cutting sequence in $\{A, B, C, D\}^{\mathbb{Z}}$ that codes a bi-infinite linear trajectory is the sequence of labels of sides hit by the trajectory.

The cutting sequence of the trajectory in the example contains (see Fig. 3.8):

$$\ldots ABBACD \ldots$$

Which sequences in $\{A, B, C, D\}^{\mathbb{Z}}$ are cutting sequences?

In an admissible sequence only certain pairs of consecutive letters (transitions) can occur. If $\theta \in [0, \pi/8]$, the transitions which can appear correspond to the arrows in the diagram given in Fig. 3.9. Denote

$$\Omega = \{A, B, C, D\}^{\mathbb{Z}}.$$

A sequence in Ω is admissible if it uses only the arrows on this diagram or these on one corresponding to another sector $\left[\frac{k\pi}{8}, \frac{(k+1)\pi}{8}\right]$ of directions.

Exercise 22.

1) Give examples of admissible sequences. Is your sequence periodic? Is there non-periodic one?

$$A \rightleftarrows D \rightleftarrows B \rightleftarrows C \circlearrowright$$

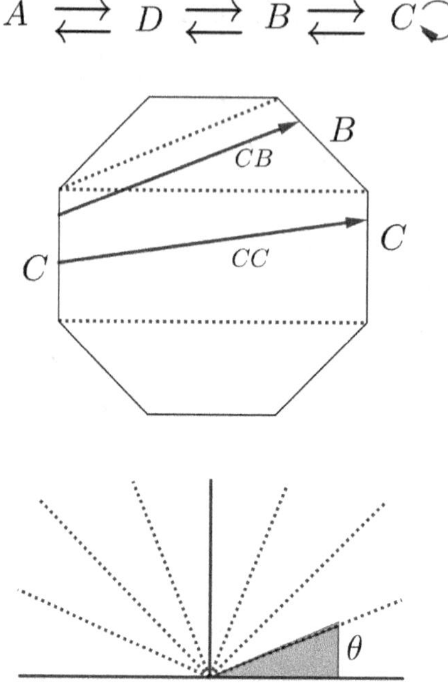

Fig. 3.9   Transitions between A, B, C, D.

2) How many sequences are there in Ω?
3) How many admissible sequences are there in Ω?
4) For which values of $k = 1, 2, 3, \ldots$, is there an k-periodic admissible sequence in Ω?

A letter is sandwiched if it is preceded and followed by the same letter, e.g. in $DBBCBAAD$ the letter C is sandwiched between two Bs.

The derived sequence of a cutting sequence is obtained by keeping only the letters that are sandwiched and erasing the other letters, e.g.

$$\ldots DADBCCBCCBDADBCBDBDBCBD \ldots,$$

is derived to

$$\ldots ABACDDC \ldots$$

A sequence in $\{A, B, C, D\}^{\mathbb{Z}}$ is derivable if it is admissible and its derived sequence is again admissible.

Theorem 3.3. [116] *Any octagon cutting sequence is infinitely derivable.*

With an additional condition, it becomes an if and only if (full combinatorial characterization of cutting sequences), analogous to the characterization of Sturmian sequences using continued fractions. For details see [106].

Exercise 23. [30] Draw a line on an infinite square grid, and record each time the line crosses a horizontal or vertical edge (see Fig. 3.10).

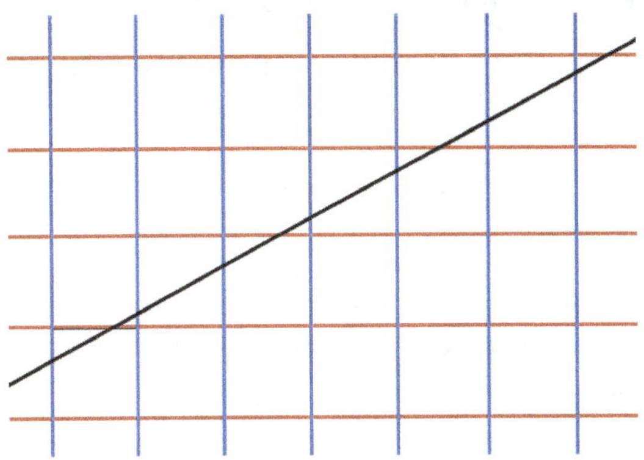

Fig. 3.10

Assume that the direction of travel along the line is always left to right and assign A to horizontal and B to vertical edges. Record it as

$$...BABBABBABBA...$$

(a) What is the slope of the line in the picture?

(b) Record these As and Bs, for several different lines. Describe any patterns you notice.

(c) What should you do if the line hits a vertex?

Exercise 24. [30] Consider a billiard trajectory in a square. Draw the unfolded trajectory. Then use this unfolding to prove that any trajectory with slope 2 yields a periodic billiard trajectory on the square. Which other slopes yield a periodic billiard trajectory?

Exercise 25. On the vertex $(0,0)$ of the square lattice $\mathbb{Z}^2$ sits a hunter, and on all other vertices of $\mathbb{Z}^2$ sits identical rabbits, which can be considered as circles of radius ϵ with their centers at these vertices. The hunter shoots at random where the bullet's trajectory is a ray l leaving from the point $(0,0)$. When will the hunter come back home with a rabbit?

In the following sections we will do the unfolding constructions for many polygonal tables.

3.3 Billiard trajectories in triangles

3.3.1 *Fagnano's problem and periodicity*

Fagnano's problem. In geometry, Fagnano's problem is an optimization problem that was first stated by Giovanni Fagnano in 1775:

Problem 3.3. *For a given acute triangle determine the inscribed triangle of minimal perimeter.*

Solution. The orthic triangle, with vertices at the base points of the altitudes of the given triangle, has the smallest perimeter of all triangles inscribed into an acute triangle, hence it is the solution of Fagnano's problem.

The following proposition gives the Fagnano billiard trajectory.

Proposition 3.1. *On an acute triangular billiard table, the triangle connecting the base points of the three altitudes is a 3-periodic billiard trajectory (Fig. 3.11).*

Proof. The quadrilateral $BPOR$ has two right angles: $\angle BRO = \angle BPO = \pi/2$. Hence it is inscribed into a circle. The angles APR and ABQ are supported by the same arc of this circle; therefore they are equal. Similarly, the angles APQ and ACR are equal. It remains to show that the angles ABQ and ACR are equal. Indeed, both complement the angle BAC to $\pi/2$. This completes the proof. $\square$

Since the distance between parallel lines does not change after reflection in a flat mirror, it follows that periodic billiard trajectories in a polygon are never isolated: an even-periodic trajectory belongs to a 1-parameter family of parallel periodic trajectories of the same period and length, and an odd-periodic one is contained in a strip consisting of trajectories whose period and length are twice as great (Fig. 3.12). The Fagnano trajectory degenerates when the triangle becomes a right one. In [39] and [57] it was shown that every right triangle also contains a periodic billiard trajectory.

Polygonal billiards are a rather difficult object of study despite the simplicity of their description. Here we give some results for triangular tables:

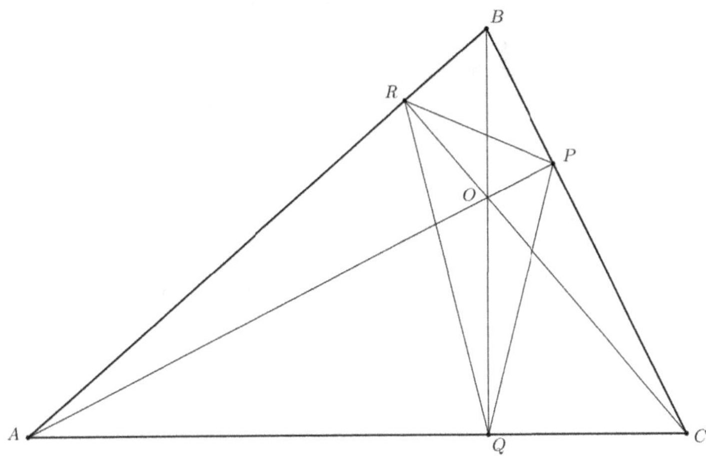

Fig. 3.11   The Fagnano billiard trajectory.

Definition 3.2. A trajectory in a triangle is called perpendicular if it meets a side of the triangle that is perpendicular to it.

Theorem 3.4. *In a right triangle, a billiard trajectory that starts on the hypotenuse in the perpendicular direction returns to this side in the same direction.*

Proof. Let the right triangle be ABC. By the reflection rule of the billiard trajectory we know that the triangles ABC, QBP, QRC, ART, shown in Fig. 3.13, are similar, since corresponding angles have the same measure. This completes the proof. $\qquad\square$

Remark 3.4. The trajectory showed in Fig. 3.13 is a 6-periodic billiard trajectory in a right triangle. Note that such a periodic trajectory, in a polygonal billiard, leaves a side in the orthogonal direction and returns in the same direction to the same side. This is the only class of polygons for which the billiard system is relatively well understood [26], [110].

Remark 3.5. [110] It is not known whether every polygon has a periodic billiard trajectory; this is unknown even for obtuse triangles. Substantial progress has recently been made by R. Schwartz, who proved that every obtuse triangle with angles not exceeding 100^0 has a periodic billiard path.

Periodic trajectories in obtuse triangles. In the case of billiards

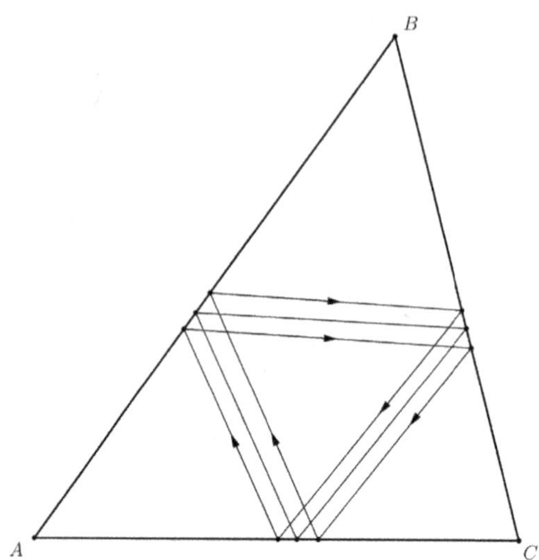

Fig. 3.12　The strip of parallel billiard trajectories.

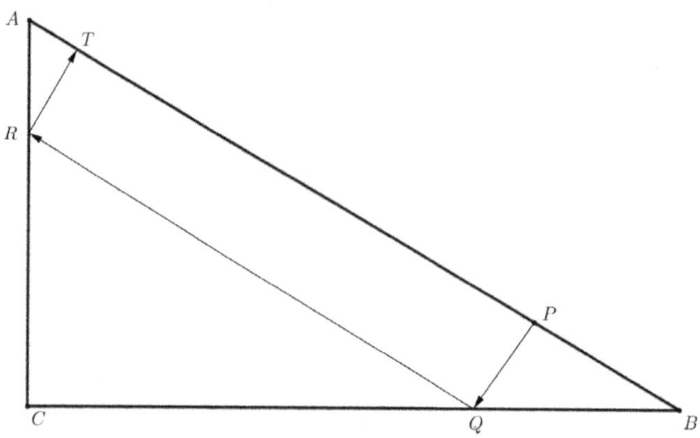

Fig. 3.13　A trajectory started perpendicularly on the hypotenuse.

in an obtuse triangle we have the following results.

Proposition 3.2. *For every n there is an obtuse triangle $\triangle(n)$ in which every periodic trajectory has more than n links.*

Proof. For a given n we choose the acute angles α and β of the triangle $\triangle(n) = \triangle ABC$ to be small enough that $\min(-\lfloor -\frac{\pi}{\alpha}\rfloor, -\lfloor -\frac{\pi}{\beta}\rfloor) > n$. $N(\alpha) = -\lfloor -\frac{\pi}{\alpha}\rfloor$ is the maximal possible number of reflections of a billiard ball in a sector of angle α. We assume α, β, and π to be rationally independent, meaning that

$$k\alpha + m\beta + l\pi = 0, \ k, m, l \in \mathbb{Z} \Rightarrow k = m = l = 0.$$

Let Γ be a periodic trajectory in $\triangle(n)$. Note that least one of its links XY has endpoints on the lateral sides AB and BC ($\angle B > \pi/2$) as seen in Fig. 3.14. Indeed, assume the billiard particle hits the base AC after each reflection on a lateral side. Therefore, any two angles of reflection in the base AC are related by $\varphi' - \varphi'' = k\alpha + m\beta$, $k, m \in \mathbb{Z}$, since adjacent angles of reflection in AC differ in modulus only by 2α or by 2β. Hence, if the ball starts under an angle φ_0 with AC, then after the first reflection in AC the angle of reflection would be $\varphi = \varphi_0 + k\alpha + m\beta$. Since Γ is a periodic trajectory, the angles φ and φ_0 differ by a number of full rotations:

$$\varphi = \varphi_0 + 2\pi s \Rightarrow k\alpha + m\beta = 2\pi s \Rightarrow k = m = s = 0, \ k, m \in \mathbb{Z}$$

by the rational independence of α, β, and π.

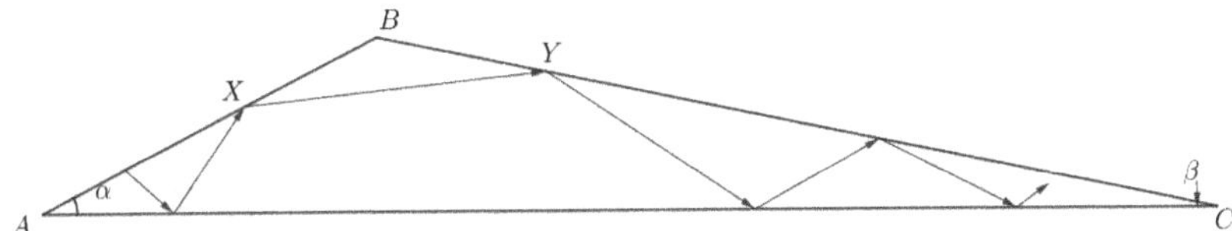

Fig. 3.14

So we have proved that Γ contains a link XY, $X \in AB$, $Y \in BC$. This link forms small an angle with the side AC that after reflection in AB and BC at X and X the particle has performed $N(\alpha)$ and $N(\beta)$ reflections inside the angles A and B, respectively. Hence Γ contains more than n links, and the proposition is proved. $\qquad\square$

In the next subsection following [41], we give some more results on periodic trajectories in right triangles.

The question of whether there exists at least one periodic trajectory in every triangle (more generally, in every polygon) is still open.

Exercise 26. Consider an arbitrary triangle $\triangle ABC$ with the biggest angle at vertex B. Let BH be the height of the triangle. We introduce an

Oxy coordinate system taking the point $H = (0,0)$, considering AC on Ox and BH on Oy (as shown in Fig. 3.15). Thus there are $a < 0$, $b > 0$ and $h > 0$ such that $A = (a,0)$, $B = (0,h)$, $C = (b,0)$. Take an arbitrary point $P_0 = (x_0, y_0)$ on the boundary of the triangle and fix $\alpha_0 \in [0, \frac{\pi}{2}]$. Consider the pair (P_0, α_0) as an initial point of the billiard trajectory. Denote by (P_m, α_m) the point of the trajectory after m reflections, i.e., $P_m = (x_m, y_m)$, $\alpha_m \in [0, \frac{\pi}{2}]$. Since (x_m, y_m) belongs to a graph of a linear function representing an edge of the triangle we can define y_m as a function of x_m.

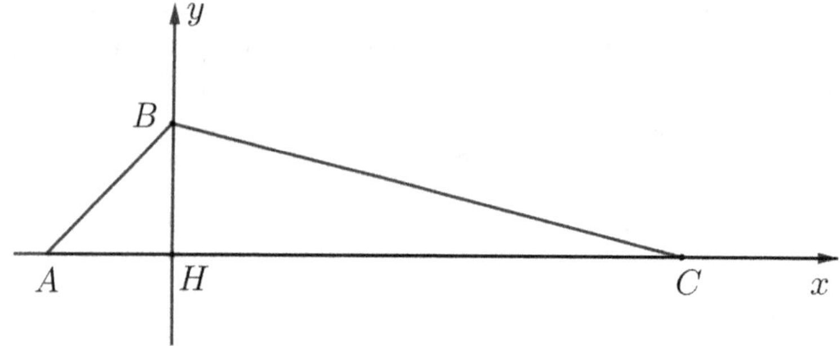

Fig. 3.15

1. Define the operator V (which depends on parameters a, b, h; moreover, it may depend on the initial point) which maps (P_0, α_0) to (P_1, α_1), i.e. $(P_1, \alpha_1) = V(P_0, \alpha_0)$.
2. Find the coordinates of (P_m, α_m), $m = 1, 2, 3$ for $y_0 = 0$ (i.e. the initial point on AC) with arbitrary α_0.
3. Find the coordinates of (P_m, α_m), $m = 1, 2, 3$ for $(x_0, y_0) \in AB$ and with arbitrary α_0.
4. Find the coordinates of (P_m, α_m), $m = 1, 2, 3$ for $(x_0, y_0) \in BC$ and with arbitrary α_0.
5. Find the conditions on (P_0, α_0) (the conditions should be given on a, b, h) to get a 3-periodic trajectory.
6. Try to find formulas for (P_m, α_m), $m = 1, 2, 3, \ldots$.

Hint. Use analytical geometry (to write equations of straight lines, to find intersection points of the lines, to find angles between straight lines) to find the operator.

3.3.2 *Trajectories that begin perpendicular to a side*

Following [26] we give the following theorem.

Theorem 3.5. *For a billiard on an arbitrary right triangle, almost all trajectories that begin perpendicular to a side are periodic.*

Proof.

Preparation to prove. By unfolding we get a representation of a trajectory as a sequence of flips of the triangle along a straight line. Fig. 3.17 represents the unfolded trajectory given in Fig. 3.16. The same sequence

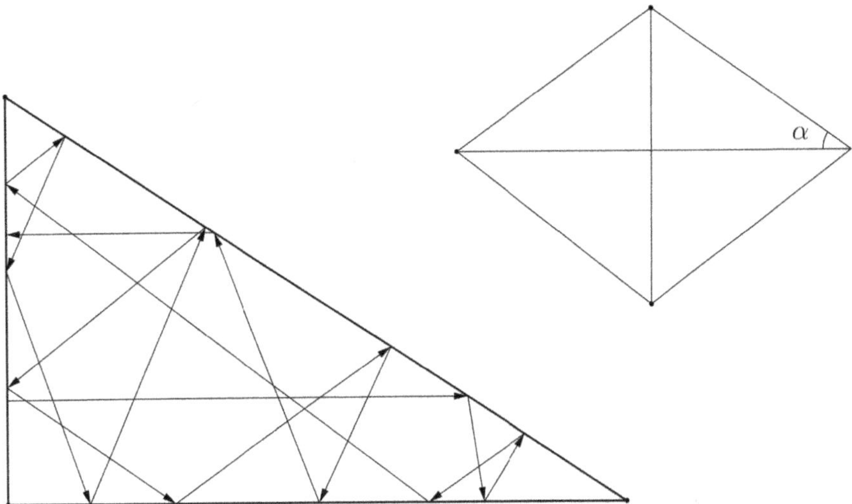

Fig. 3.16   A 15-segment periodic trajectory within a right triangle starting perpendicular to the short leg. In this case the ratio of the short to long leg is 1 : 1.42 (left). The rhombus corresponding to this triangle has $\tan \alpha = 1/1.42$.

of flips (unfolds) can be done to all trajectories in a band about the chosen trajectory. The upper and lower limits of the band are determined by where the vertices of the flipped triangles land: The upper limit is determined by the lowest vertex above the chosen trajectory, and the lower limit by the highest vertex below. One of these vertices (in this case the one above) is the right-angle vertex. If we look at a trajectory which is as much above this vertex as the original trajectory was below it, we see that the sequence of flips is essentially the same (see Fig. 3.18). In fact, as billiard trajectories

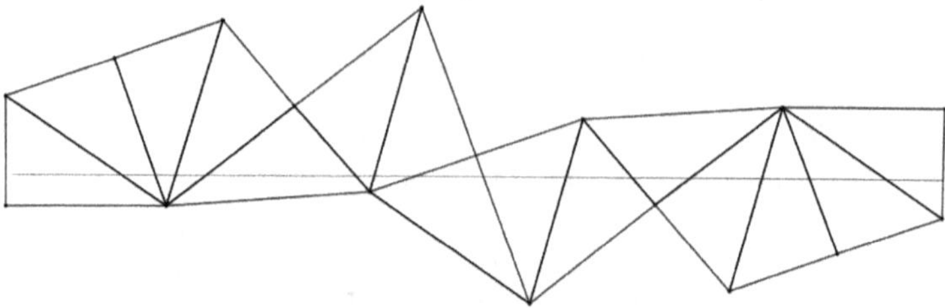

Fig. 3.17   Unfolded trajectory from the right triangle given in Fig. 3.16.

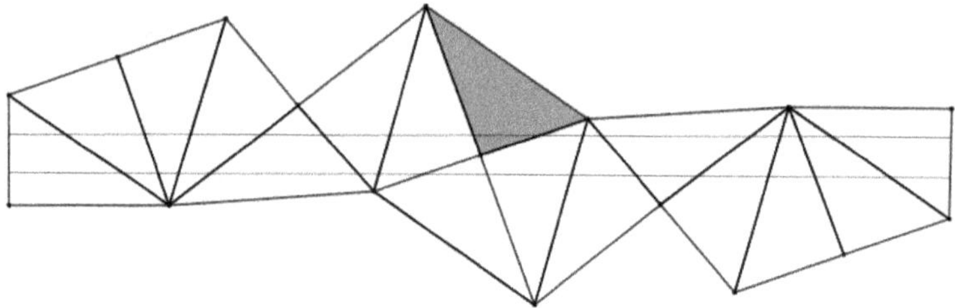

Fig. 3.18   The trajectory of Fig. 3.17 along with its equivalent around the point of symmetry.

inside the triangle, these two are identical. The right-angle vertex is a point of rotational symmetry in Fig. 3.18. This suggests that the right-angle vertex is a "removable singularity". We can remove it by using not the right triangle, but a rhombus consisting of four copies of the triangle as our basic shape, as shown in Fig. 3.16.

Thus we can work with rhombi, and consider trajectories which begin perpendicular to one of the diagonals of the rhombus. Since we follow the flips of the rhombus along a straight line trajectory, each flip in effect rotates the rhombus by an amount equal to one of the two interior angles of the rhombus. It is convenient to describe the rotations in terms of one angle only.[3]

Let α be one of the angles of the right triangle, then as the rhombus flips along a trajectory, its orientation increases or decreases by 2α with each flip. Therefore, we label the rhombi with integers, according to the total number of, say clockwise, rotations by 2α, for each, as shown in Fig. 3.19.

---

[3] A clockwise rotation by one interior angle is equivalent to a counterclockwise rotation by the other.

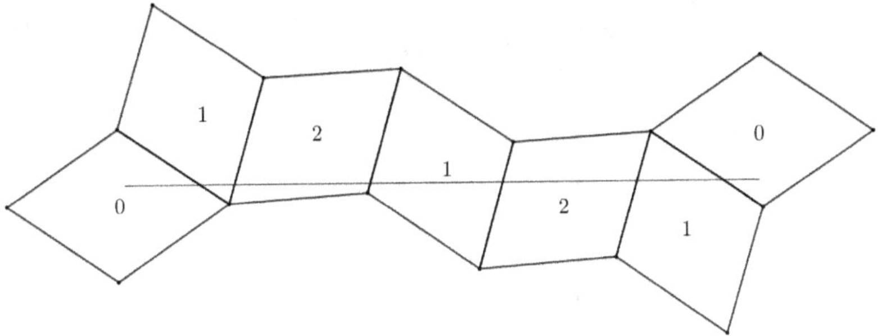

Fig. 3.19   The trajectory of Fig. 3.16, as a sequence of rhombus flips. Here the numbers 0, 1, 2 refer to unique rhombus orientations.

This suggests a new representation of a trajectory as a sequence of line segments in a set of labeled rhombi. An example is shown in Fig. 3.20.

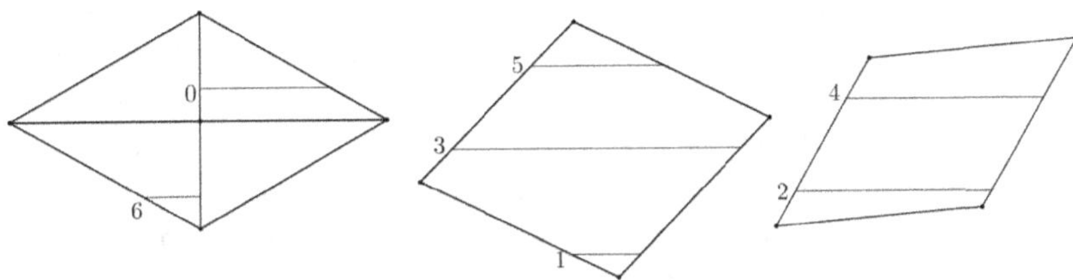

Fig. 3.20   The rhombus 0 left, rhombus 1 middle and rhombus 2 right. These are three unique rhombus orientations of Fig. 3.19 with sequential line segments of the trajectory indicated.

The segments of the trajectory are numbered as they enter each rhombus from the left. When a segment exits a rhombus on the right, it either advances to the next rhombus, or goes back to the previous one, reentering at the corresponding point on the edge parallel to the edge it has just left.

A generic n-th rhombus is shown in Fig. 3.21.

The two sides through which a horizontal, right-moving trajectory can exit are labeled "forward" and "back", according to whether the trajectory is to go forward to the next rhombus or back to the previous one. The forward edge is always the leading (clockwise) ray for the angle 2α, and the backward edge is the trailing (counterclockwise) ray. If α is a rational multiple of π, then there are only finitely many distinct orientations for

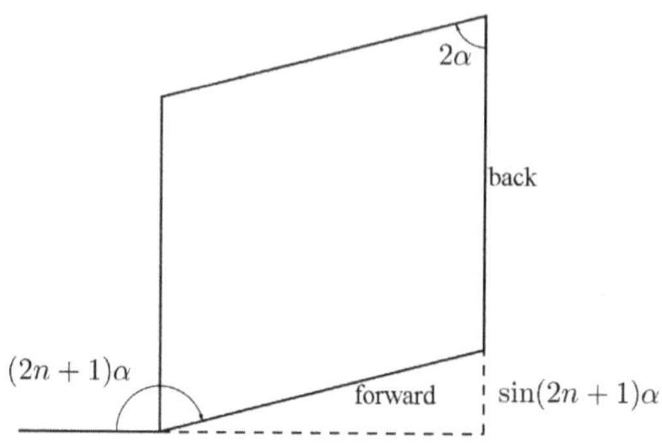

Fig. 3.21

the labeled rhombi, while for an irrational angle α, there are infinitely many orientations. It is clear that the forward edge in Fig. 3.21 is nearly horizontal.

Since the forward edge of rhombus 0 is set at angle α with respect to the x-axis, and each subsequent forward edge is a clockwise rotation by 2α, taking the rhombus to have sides of unit length, we see that the vertical width of the forward edge of rhombus n is $\|\sin[(2n+1)\alpha]\|$. We note[4] that if α is an irrational multiple of π, then $\|\sin[(2n+1)\alpha]\| \neq 0$, but takes values that are arbitrarily close to 0 as n ranges over the positive integers. To prove that almost all trajectories are periodic, we use the fact that the vertical width of the forward edge of the n-th rhombus is arbitrarily small for certain values of n.

Recall now that each periodic trajectory actually belongs to a band of trajectories that follow the same pattern of flips. This carries over to the new representation given in Fig. 3.22.

The width of the band is determined by the vertices P_1, and P_2, which separate the forward and backward edges of rhombus 1 and rhombus 2, respectively. Note that any periodic trajectory can be widened into such a band.

Fig. 3.22 can be reformed as shown in Fig. 3.23.

This "pinwheel" configuration of the rhombi makes it easier to see the progression of the band. However, the analysis is based on considering each rhombus separately.

---

[4]Using Kronecker's theorem: the set of fractional parts of the integer multiples of any irrational number is dense in the unit interval [55].

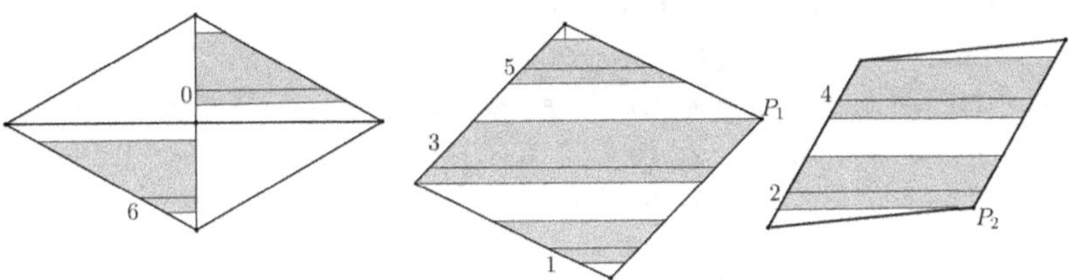

Fig. 3.22 The band of trajectories containing the trajectory of Fig. 3.20 and bounded by points P_1 and P_2.

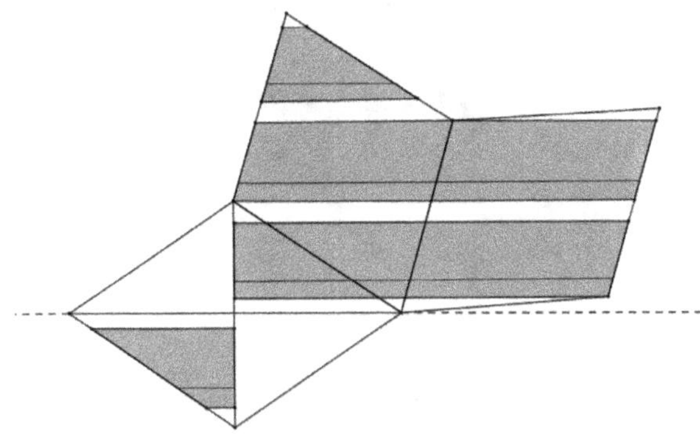

Fig. 3.23 The rhombi of Fig. 3.22 arranged in "pinwheel" configuration.

Because the rule that ties trajectories from one rhombus to another simply amounts to an affine translation, the width of the band remains constant. Furthermore, the segments of the band, such as segments 3 and 5 in rhombus 1, do not overlap. Indeed, if segments h and k overlapped in one rhombus, with $h < k$, then segments $h - 1$ and $k - 1$ would have overlapped in the preceding rhombus. This argument would fail if $h = 0$, since there is no "1" segment, but since segment 0 starts in rhombus 0, which is the one that is symmetrically oriented, there can be no segment that overlaps with it: Any segment that reenters rhombus 0 does so in its lower half, and stops at right angles at the diagonal from which segment 0 began. This non-overlapping of segments holds between bands as well, for the same reason. That is, let B_1, and B_2 be two bands of trajectories, each starting at right angles to the vertical diagonal of rhombus 0. Then no segment of band B_1 overlaps with any segment of band B_2. The non-overlapping of bands and the constancy of their width will be also used to prove that almost all trajectories are periodic. Now we are ready to prove the theorem.

Proof of the theorem: Consider what happens to a "beam" consisting of all trajectories perpendicular to the vertical diagonal in the top half of rhombus 0. It is transmitted as a whole to rhombus 1, as shown in Fig. 3.24.

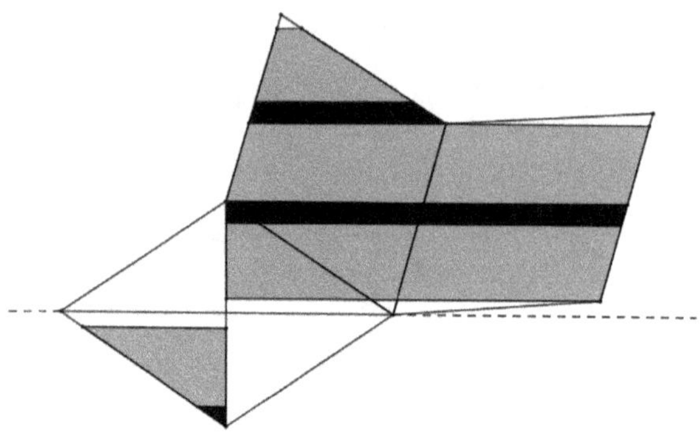

Fig. 3.24 The two bands of trajectories which involve only the first three rhombus orientations and account for approximately 87 percent of all trajectories beginning perpendicular to the short leg.

In that example, the beam is again transmitted as a whole to rhombus 2, but that rhombus splits the beam, sending the lower part of it on to rhombus 3, and the higher part back to rhombus 1. We observe that the latter subbeam is "trapped" in rhombi 1, 2, and 0: The only entrance to rhombus 3 is through the forward edge of rhombus 2, but that edge is fully filled, and beams, like bands, cannot overlap. The trapped subbeam is split again, this time by the vertex in rhombus 1, but both of these subbeams finally wind up in rhombus 0, where they terminate. Thus the top part of the initial beam is trapped in the first few rhombi while the bottom part goes further. This happens whenever there is an integer m, depending on α, for which

$$0 < \sin[(2m + l)\alpha] < \sin \alpha$$

but $\sin \alpha < \sin[(2k + 1)\alpha]$ for $0 < k < m$. The initial beam advances as a whole to rhombus m, where it is split, because, the condition is equivalent to the double inequality $0 \leq \pi - (2m + 1)\alpha < \alpha$, but such an integer m exists if and only if the integer part of $\frac{\pi}{\alpha}$ is odd.

The bottom, "transmitted" subbeam fills up the forward edge leading to rhombus $m+1$, while the top subbeam is trapped in rhombi $0 - m$. The trapped subbeam may be split again by the vertices of rhombi $1, 2, \ldots, m-1$, but all subbeams eventually wind up in rhombus 0, where they terminate. Because beams of constant width cannot continue indefinitely in a finite region without overlapping. In other words, due to the non-overlapping condition, each vertex separating the forward and backward edges of a rhombus can act as a "beam splitter" at most once. Hence the described "trapped" beam splits into at most m subbeams.

In the Fig. 3.25 the continuation of the transmitted beams in rhombi 3, 4, and 5 is shown.

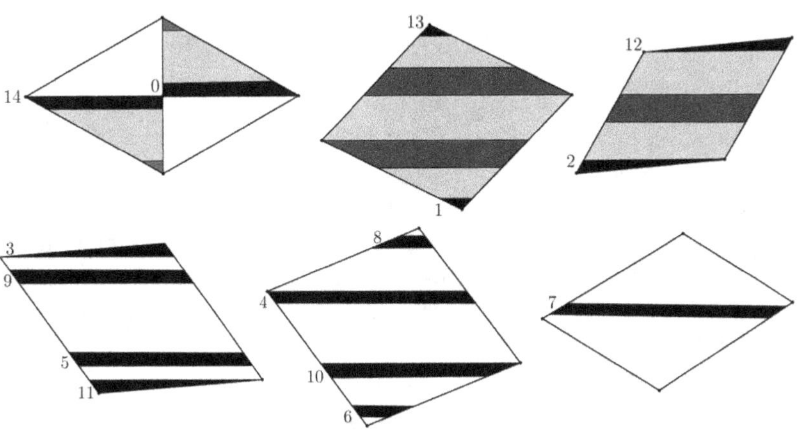

Fig. 3.25 Rhombus: 0, 1, 2, 3, 4, 5 (from the top left). The three bands of trajectories which involve only the first six rhombus orientations and account for approximately 97.7 percent of all trajectories which begin perpendicular to the short leg.

Only a very thin sliver (less than 2.5 percent) of the original beam is sent on to rhombus 6; the rest, consisting of three subbeams, two "trapped" beams and most of the "transmitted" beam, is confined to rhombi 0-5. For the same picture in the pinwheel configuration see Fig. 3.26.

We know that, if a right triangle has angles that are rational multiples of π, then there are only a finite number of rhombi. Therefore, the beam, which is split at most a finite number of times, cannot contain a subbeam

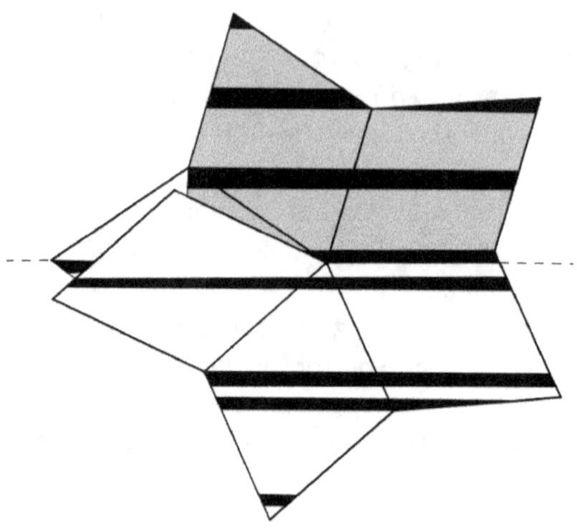

Fig. 3.26   The pinwheel configuration of Fig. 3.25

that continues indefinitely. This suffices to prove that all trajectories perpendicular to a side, except those that hit a vertex, are periodic.

But what about the irrational case? In general, when a beam enters rhombus n, one of following three things happens: It can exit entirely through the forward edge and go on to rhombus $n + 1$, it can exit entirely through the backward edge and be sent back to rhombus $n - 1$, or it can be split in two. But since there are integers n for which the vertical width of the forward edge of the n-th rhombus is arbitrarily small, for such an n, an arbitrarily large fraction of the initial beam will be trapped in rhombi $0, 1, \ldots, n$ in the form of at most n subbeams. The trajectories comprising these beams (excluding those that hit vertices) are necessarily periodic.

If there is a non-periodic trajectory, it is contained in a beam of increasingly narrow width, and such trajectories constitute a set of measure zero. Thus we have proved that almost all trajectories perpendicular to a side of a right triangle with irrational angles are periodic.   □

Now following [118], we give "perpendicular" trajectories in obtuse triangles of special shapes. Consider an obtuse triangle where its acute angles α and β are related in a certain way (see Fig. 3.27). A 6-periodic trajectory in this triangle has the property that it leaves one side (AC) perpendicular to it and falls on the other side (BC) also perpendicular to it. A trajectory leaving one side orthogonally is naturally called a "perpendicular" trajectory.

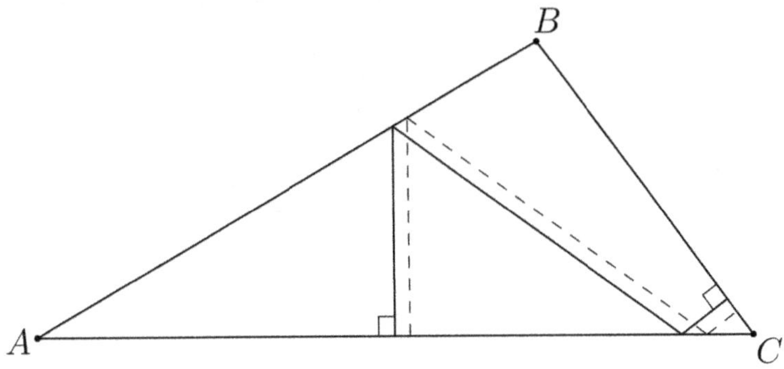

Fig. 3.27

Proposition 3.3. *If the acute angles α and β of a triangle ABC satisfy the conditions $k\alpha = m\beta < \frac{\pi}{2}$, where $k, m \in \mathbb{N}$ are natural numbers, then ABC contains a perpendicular $2(k+m)$-periodic trajectory.*

Proof.   To prove this we use the unfolding method described in the previous section. We construct the chain of triangles:

$$A_{-m}B_{-m}C_0, \ldots A_{-1}B_{-1}C_0, A_0B_0C_0 = ABC, A_0B_1C_1, \ldots, A_0B_{k-1}C_{k-1},$$
$$(3.1)$$

in which every two adjacent triangles have a common side and are positioned symmetrically with respect to this side (Fig. 3.28). For this, we mirror-reflect the angle $B_0A_0C_0$ $k-1$ times counterclockwise around the vertex A_0, and reflect the angle $B_0C_0A_0$ m times counterclockwise around the vertex C_0. One of the sides of the triangle $A_0B_{k-1}C_{k-1}$, (denoted by A_0X), forms the angle $k\alpha$ with A_0C_0. Similarly, one of the sides of $A_{-m}B_{-m}C_0$, denoted by C_0Y, forms the angle $m\beta$ with C_0A_0. Since $k\alpha = m\beta$, we know that A_0X and C_0Y are parallel. Using the conditions $k\alpha < \pi/2$ and $m\beta < \pi/2$, we can draw MN which is perpendicular to A_0X and C_0Y, separating the points B_0 and B_{-1}. The segment MN is entirely contained in the corridor Φ formed by the triangles (3.1). Note that the closed polygonal line obtained by traversing MN twice is a periodic trajectory in Φ. Assume MN intersects the sides of the triangles given in (3.1) at the points

$$N = M_{-m}, M_{-m+1}, \ldots, M_{-1}, M_0, M_1, \ldots, M_{k-1}, M_k = M.$$

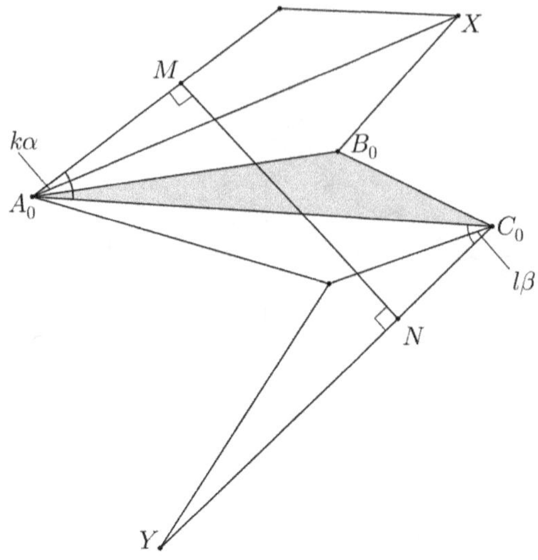

Fig. 3.28

We now 'fold' the corridor Φ like an accordion so that we obtain $k + m$ triangular layers positioned above the triangle $A_0B_0C_0$ (the triangle $A_0B_0C_0$ itself is one such layer). Under this stacking the segments $M_{i-1}M_i$, $(m + 1 \leq i \leq k)$ become the links of a perpendicular periodic trajectory γ in $A_0B_0C_0$, which has $2(k + m)$ links (Fig. 3.29). $\qquad\square$

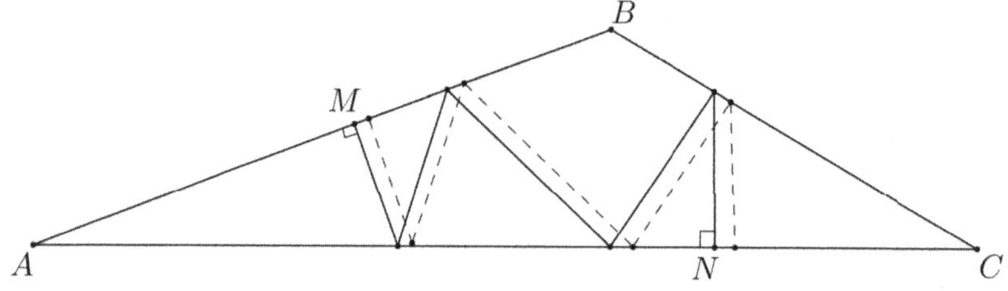

Fig. 3.29

In the particular case $k = m = 1$ we obtain a bunch of the simplest, 4-periodic trajectories in an isosceles triangle. By 'stacking' the isosceles triangle along its altitude we obtain a bunch of 6-periodic trajectories in a right triangle. So we have the following,

Proposition 3.4. *In any isosceles and any right triangle there are bunches of perpendicular trajectories.*

There are also perpendicular periodic trajectories in obtuse triangles whose acute angles satisfy the relations

$$\alpha + k\beta = \frac{\pi}{2}, \quad k \in \mathbb{N}; \quad 2\alpha + \beta > \frac{\pi}{2}.$$

The construction of periodic trajectories in them is sketched in Fig. 3.30.

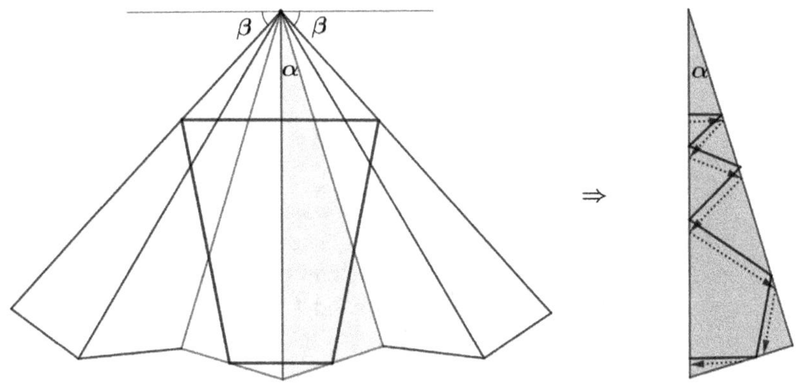

Fig. 3.30

3.3.3 *Periodic trajectories cover the triangle*

By Proposition 3.2 we know some triangles have periodic trajectories with length bounded from below, i.e. for any n there exist some triangles where every periodic trajectory has more than n links.

Definition 3.3. A periodic trajectory in a triangle is stable if for any small perturbation of the triangle, the triangle obtained contains a periodic trajectory close to the initial one.

One says that two trajectories are close if they have the same number of reflection points and the corresponding reflection points lie on the same edge of the triangle close to each other.

To construct a periodic trajectory in a right triangle one takes two steps. First, reflect the triangle on its legs to obtain a rhombus thats diagonals are composed of two legs of the triangle (see Fig. 3.31).

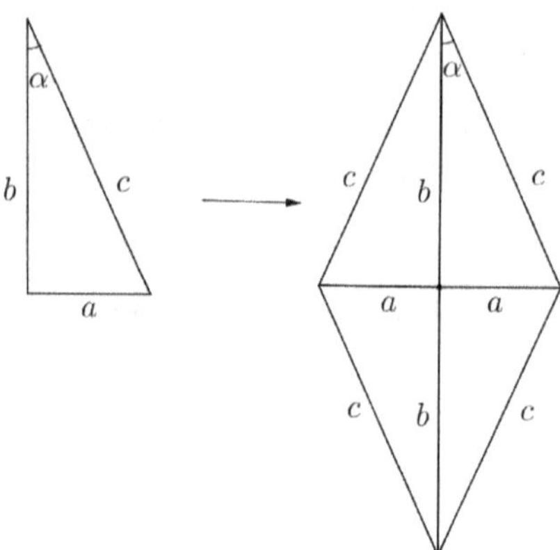

Fig. 3.31   Construction of a rhombus from the right triangle.

The vertices of the acute angles of the rhombus will be called its extremities. Second, reflect the rhombus on its sides either following a straight line assumed to be the unfolding of a periodic trajectory, or constructing a corridor of rhombuses in such a way that the last rhombus is parallel to the initial one. After that, join two identical points in these rhombuses by a segment which, if entirely contained in the corridor, is the unfolding of a periodic trajectory.

In the theory of mathematical billiards in a polygon, the trajectories that get into the vertex of an angle are considered singular and are said to end at this vertex. But the right angle, having the form π/n, is a removable singularity: if two close trajectories hit different sides of the angle their unfoldings in the rhombus remain close and therefore the trajectories themselves remain close. It is possible to make two close trajectories remain close after leaving the angle even if one of them gets into the vertex. To achieve this, we assume that a particle getting into the vertex is merely reflected back. In other words, the unfolding of a trajectory that gets into the vertex of the right angle continues straight forward in the rhombus.

Let us denote the smaller acute angle of the triangle by α. In this case the acute angle of the rhombus is equal to 2α. Since the rhombus is symmetric, its reflection on a side is equivalent to a rotation around its extremity by an angle $+2\alpha$ or -2α, depending on the direction of rotation.

Theorem 3.6. *Over every point of a right triangle passes a periodic trajectory.*

Proof. We use the following construction (see Fig. 3.32):

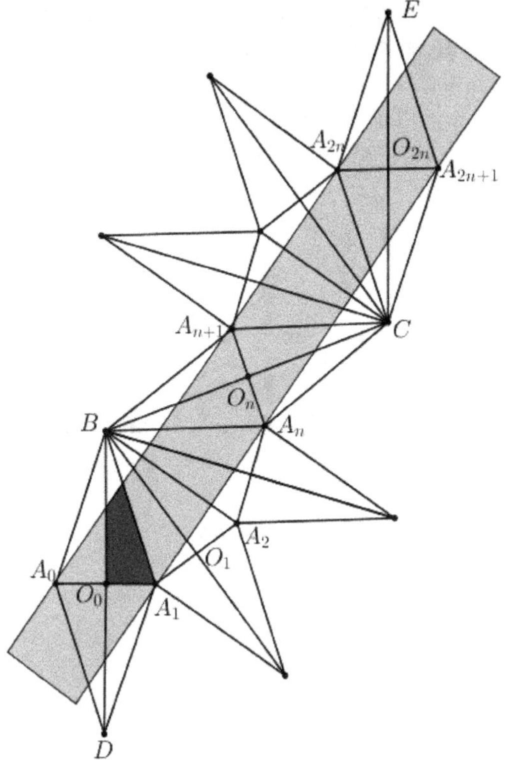

Fig. 3.32

1. reflect the initial rhombus on its side n times so as to make it turn counterclockwise by $2\alpha n$ around the vertex B;
2. reflect n more times so that it turns clockwise, by $-2\alpha n$, around the vertex C,

where $n = 1, 2, \ldots, N$ with $N = \lceil \frac{\pi}{2\alpha} \rceil$. After such reflections the first and the last rhombuses will be parallel to each other (see the vertical rhombuses in Fig. 3.32). Moreover, the corridor of rhombuses shown in Fig. 3.32 has a center of symmetry. It coincides with the center O_n of the $n-th$ rhombus, i.e., the rhombus on which we change the direction of rotation.

We trace the maximally wide bundle of segments parallel to the line $O_0 O_n O_{2n}$ and entirely contained in the rhombuses constructed. This bundle is called a strip. Note that all the segments of the strip (except its borders) are unfoldings of periodic trajectories in $\triangle B O_0 A_1$ (see Fig. 3.32). Such a trajectory is called an S-trajectory. Clearly, every n yields its own strip. Symmetry arguments show that $O_0 O_{2k-1}$ is perpendicular to $B A_k$ while $O_0 O_{2k}$ is perpendicular to $B O_k$ for every $k \geq 1$. This is used to prove the following

Proposition 3.5.

    i) S-trajectories are perpendicular.

    ii) The n-th strip is inclined at an angle $n\alpha$ with respect to the line $O_0 A_1$.

Proof. i) The trajectories of the n-th strip are parallel to $O_0 O_n$. Therefore they meet either the longer leg $B A_k$ or the hypotenuse $B O_k$ of the triangle that is perpendicular to it.

ii) For an even n, i.e., $n = 2k$, the sides of the angle $\angle A_1 O_0 O_n$ are perpendicular to the sides of the angle $\angle O_0 B O_k$. Therefore

$$\angle A_1 O_0 O_n = \angle O_0 B O_k = n\alpha. \tag{3.2}$$

Similarly, for an odd n, i.e., $n = 2k - 1$, the sides of the angle $\angle A_1 O_0 O_n$ are perpendicular to the sides of the angle $\angle O_0 B A_k$ and we have again (3.2).

Let us construct two bundles of periodic trajectories and show that they cover the triangle entirely. Namely, we show that the strips corresponding to the two maximal values of n ($n = N - 1$ and $n = N$) cover the triangle $B O_0 A_1$ entirely.

The Fig. 3.32 represents the first strip, with $n = N - 1$, which is bounded by the lines $A_0 A_{n+1} A_{2n}$ and $A_1 A_n A_{2n+1}$. Indeed, it cannot be wider, because otherwise its boundary would have passed either further to the left than A_0 or further to the right than A_{2n+1}. And it is entirely contained in the rhombuses because the polygons $B A_0 A_1 \ldots A_n A_{n+1}$ and $C A_n A_{n+1} \ldots A_{2n} A_{2n+1}$ are convex. The part of the triangle $B O_0 A_1$ covered by this strip is shaded black in Fig. 3.32.

The second strip is shown in Fig. 3.33, which corresponds to $n = N$. Similarly one obtains that it is bounded by the lines BE and DC. The part that it covers is shaded black in Fig. 3.33. Therefore the lines $A_0 A_{n+1}$, $O_0 O_n$, and $A_1 A_n$ in Fig. 3.32 coincide with the lines $A_0 A_n$, $O_0 O_{n-1}$, and

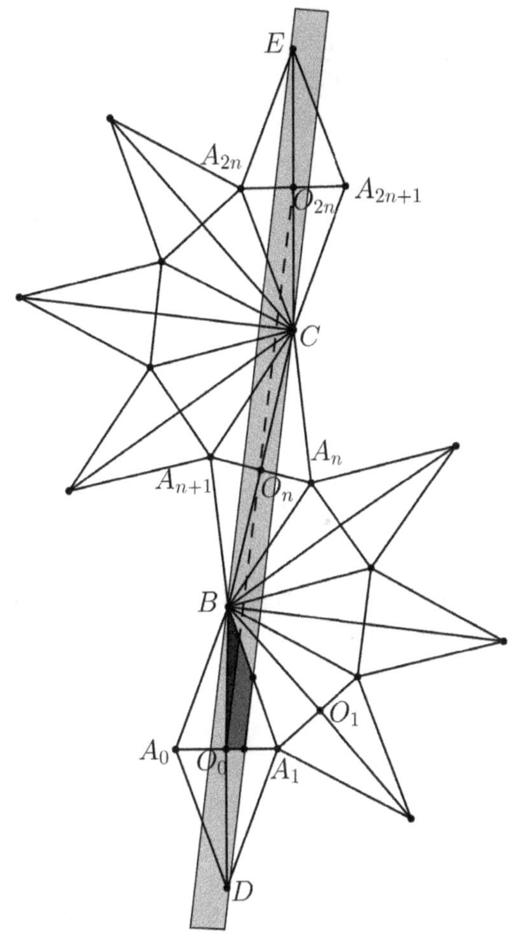

Fig. 3.33

A_1A_{n-1}, respectively in Fig. 3.33. Now we show that the two black parts cover the triangle entirely (see Fig. 3.34, where $n = N$). Note that the triangle BO_0A_1 is entirely covered if and only if the lines CD and A_0A_n intersect outside it, as in Fig. 3.34 (left), and not inside it, as in Fig. 3.34 (right).

Now, the lines A_0A_n and DC are inclined at angles $(N-1)\alpha$ and $N\alpha$ with respect to the line O_0A_1. Therefore the angle between them equals α, i.e., $\angle A_0ID = \alpha$. Since $\angle A_0BD = \alpha$ as well, it follows that the points D, A_0, B, and I lie on the same circle (see Fig. 3.35).

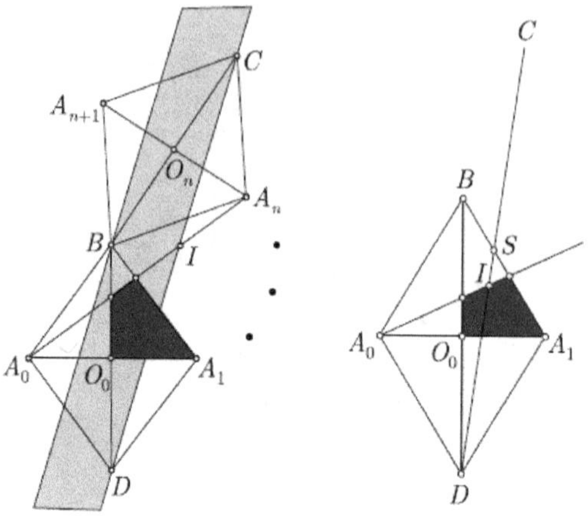

Fig. 3.34

The angle $\angle DA_0B$ is the obtuse angle of the rhombus. Therefore the circle circumscribed around $\triangle DA_0B$ lies entirely outside the rhombus. The point I is on this circle, therefore it lies outside the rhombus and in particular outside $\triangle BO_0A_1$. Hence the triangle is entirely covered by the two strips considered. $\square$

3.3.4 *Mirror periodic trajectories*

Now we shall describe so called "mirror" periodic trajectories. In Fig. 3.36 the vertical rhombus $\mathbb{A}$ rotates first by 2α around the upper extremity, then by -2α around the lower one, thus giving a new vertical rhombus denoted by $\mathbb{B}$. Consequently rotate the rhombus $\mathbb{B}$ by -2α about the lower extremity, then by 2α around the upper one. Denote the obtained vertical rhombus by $\mathbb{C}$, which is the image of the initial rhombus $\mathbb{A}$ under a horizontal translation. The same two operations are further applied alternately to the rhombuses $\mathbb{C}$, $\mathbb{D}$, $\mathbb{E}$, etc. Thus we obtain a corridor of rhombuses. The shorter diagonals of the vertical rhombuses are thought of as mirrors, and this is why we use the name "mirror periodic trajectories".

Hence a beam launched from a point of the very first mirror to the identical point of an upper mirror will move periodically in the corridor between the mirrors. If entirely contained in the union of rhombuses, it will form an unfolding of a periodic trajectory.

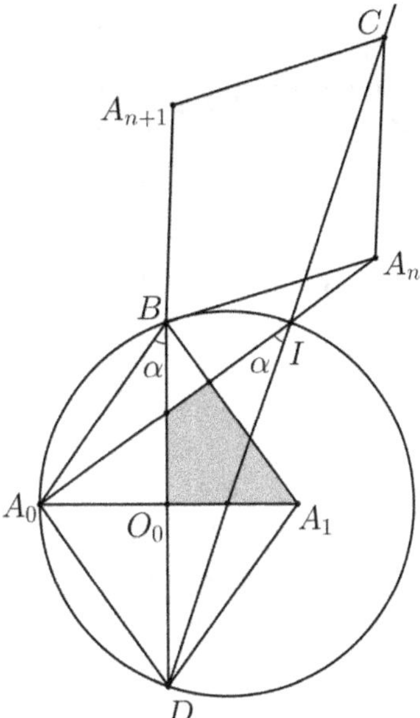

Fig. 3.35

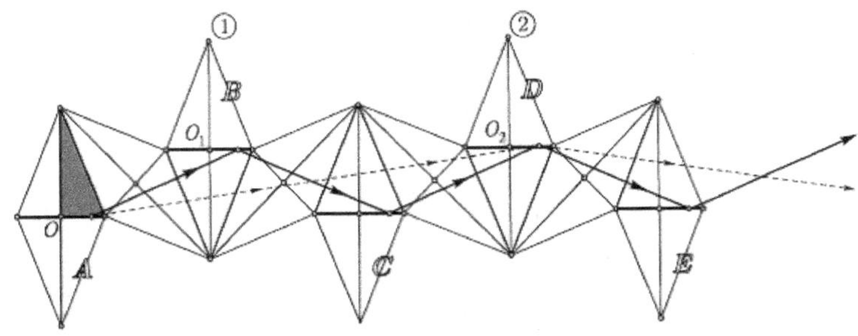

Fig. 3.36

The upper rhombuses are numbered by numbers $1, 2, 3$, etc. (Fig. 3.36). Denote the center of the very first rhombus by O and the center of the n-th upper rhombus by O_n for every n. The beginning of every beam is parallel to the line OO_n for some n. The number n uniquely identifies the type of the trajectory. A trajectory is considered to be of type n if its beginning is parallel to OO_n.

Theorem 3.7.

> *a) If $\alpha \le \frac{\pi}{6}$ then there exist mirror trajectories of type n for all n.*
>
> *b) If $\alpha > \frac{\pi}{6}$ then there exist mirror trajectories of type n only when*

$$n < -\frac{2\sin^2 \alpha}{4\cos^2 \alpha - 3} + \frac{1}{2}.$$

Proof.   a) Consider the strip bounded by two horizontal lines, the first containing the upper mirrors, the second the lower ones (see Fig. 3.37).

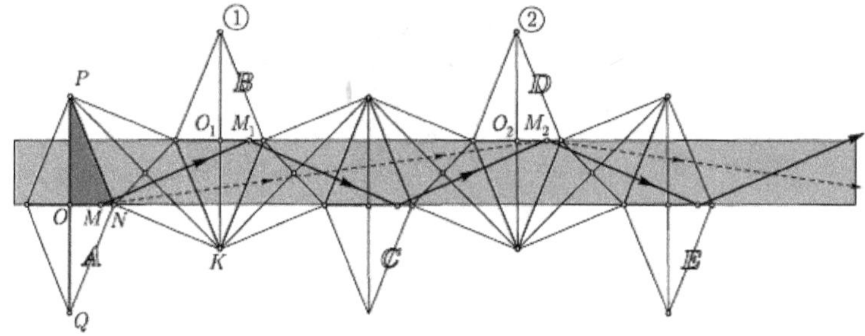

Fig. 3.37

The beams for all n are contained in this strip. It is clear that the point Q lies below the strip. Moreover, K also lies below the strip because

$$\angle ONP + \angle PNK \ge \frac{\pi}{3} + \frac{2\pi}{3} = \pi.$$

Similarly, all the other lower extremities of rhombi lie below the strip and their upper extremities, above the strip. Thus the strip and therefore the beams for all n are entirely contained in the corridor of rhombi. Therefore, it is proved that for $\alpha \le \frac{\pi}{6}$, there exist mirror trajectories of all types.

b) For every n we will find the angle φ_n at which the trajectories of the n-th type are inclined with respect to the line ON as shown in Fig. 3.38. This angle is equal to the angle between the horizontal line ON and the segment OO_n. We know that $\angle O_1 ON = \alpha$ because the sides of the angle $O_1 ON$ are perpendicular to those of the angle $\angle OPN$. Consequently, $\varphi_1 = \alpha$. The projection of the segment OO_n on the vertical axis is equal to that of OO_1, whereas its projection on the horizontal axis is $2n - 1$ times as big as that of OO_1. Hence $\tan \varphi_n = \frac{\tan \varphi_1}{2n-1}$, and therefore

$$\varphi_n = \arctan\left(\frac{\tan \alpha}{2n - 1}\right). \tag{3.3}$$

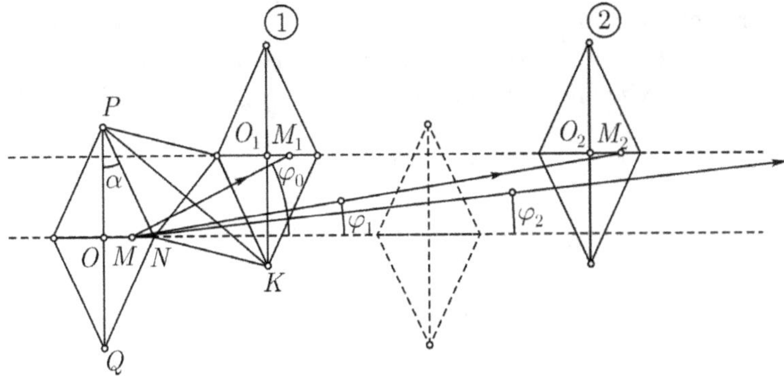

Fig. 3.38

Assume that the smaller acute angle of the triangle is equal to $\frac{\pi}{6} + \epsilon$, $0 < \epsilon \leq \frac{\pi}{12}$ (see Fig. 3.39).

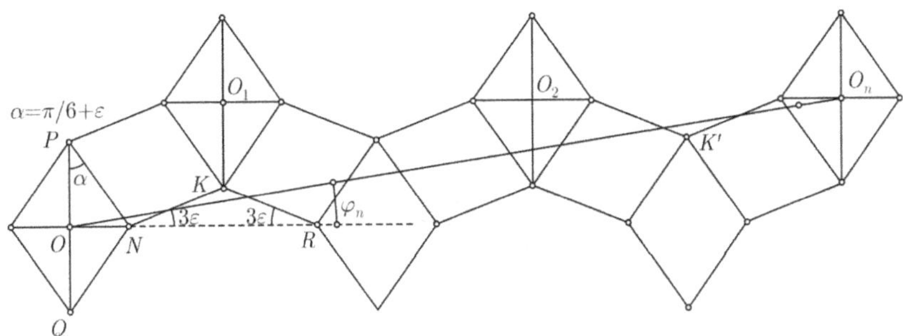

Fig. 3.39

Then the vertex K lies above the line ON. Since $\angle ONP + \angle PNK = \pi - 3\epsilon$, the angles $\angle RNK$ and $\angle NRK$ equal ϵ. Now we show that mirror trajectories of type n exist if and only if the line OO_n passes above K. Indeed, if it passes above K, it follows from symmetry arguments that it passes below K' (see Fig. 3.39). Therefore it passes below all the upper extremities of the rhombi and above all the lower extremities. Consequently it is contained in the corridor of rhombi and is surrounded by a bundle of trajectory unfoldings. Inversely, if the line OO_n passes through or below K and therefore through or above K', any line parallel to it will pass either below K or above K' and will not be an unfolding of a trajectory. Thus mirror trajectories of type n exist if and only if $\varphi_n > \angle KON$, where φ_n is determined by formula (3.3). Denoting the shorter leg of the triangle by a, the longer one by b, and the hypotenuse by c we get

$$\angle KON = \arctan\left(\frac{c\sin(3\epsilon)}{a + c\cos(3\epsilon)}\right) = \arctan\left(\frac{\sin(3(\alpha - \frac{\pi}{6}))}{\frac{a}{c} + \cos(3(\alpha - \frac{\pi}{6}))}\right)$$

$$= \arctan\left(\frac{\sin(3\alpha - \frac{\pi}{2})}{\sin\alpha + \cos(3\alpha - \frac{\pi}{2})}\right) = -\arctan\left(\frac{\cos(3\alpha)}{\sin\alpha + \cos(3\alpha)}\right).$$

Hence,

$$\varphi_n > \angle KON \quad \Leftrightarrow \quad \arctan\left(\frac{\tan\alpha}{2n - 1}\right) > -\arctan\left(\frac{\cos(3\alpha)}{\sin\alpha + \sin(3\alpha)}\right)$$

$$\Leftrightarrow \quad \frac{\tan\alpha}{2n - 1} > -\frac{\cos(3\alpha)}{\sin\alpha + \sin(3\alpha)}$$

$$\Leftrightarrow \quad 2n - 1 < -\tan\alpha \cdot \frac{\sin\alpha + \sin(3\alpha)}{\cos(3\alpha)}$$

$$\Leftrightarrow \quad n < -\frac{1}{2}\left(\tan\alpha \cdot \frac{\sin\alpha + \sin(3\alpha)}{\cos(3\alpha)} - 1\right)$$

$$= -\frac{2\sin^2\alpha}{4\cos^2\alpha - 3} + \frac{1}{2}.$$

Thus, the total number of trajectories is

$$k = \left\lfloor -\frac{2\sin^2\alpha}{4\cos^2\alpha - 3} + \frac{1}{2}\right\rfloor.$$

$\square$

The following proposition is useful.

Proposition 3.6. *The period of a mirror trajectory of the n-th type consists of* $10n - 4$ *links.*

Proof. Assume the beam is launched from a point M. For all n denote by M_n the corresponding point of the n-th upper mirror (see Fig. 3.38). In the unfolding, a period of a trajectory of type n is represented by the segment MM_n. Therefore we have to prove that the sides and diagonals of the rhombi divide this segment into $10n - 4$ parts. It suffices to note that the segment MM_1 is indeed divided into $10 \cdot 1 - 6 = 4$ parts and that we add 10 new parts by changing MM_n to MM_{n+1}.

$\square$

3.3.5 *Instability of periodic trajectories*

In this section we prove the following theorem.

Theorem 3.8.

> A. *Mirror periodic trajectories are unstable.*
> B. *Perpendicular periodic trajectories are unstable.*
> C. *S-trajectories are unstable.*

Proof. A. Instability of the mirror trajectories. In [39] a criterion of instability is given. Let us recall the criterion first. Let a_1, a_2, a_3, etc be the sides that the particle meets. Consider the alternating sum of these sides, i.e., the expression $a_1 - a_2 + a_3 - \cdots \pm a_N$. This sum is taken over one period if the length of the period is even, and over two periods if this length is odd. The criterion asserts that the trajectory is stable if and only if the alternating sum vanishes:

$$a_1 - a_2 + a_3 - \cdots \pm a_N = 0$$

i.e., if each letter occurs with a $+$ as many times as with a $-$. The sequence $a_1 a_2 \ldots a_N$ is called the code word of the trajectory. Denote the smaller leg

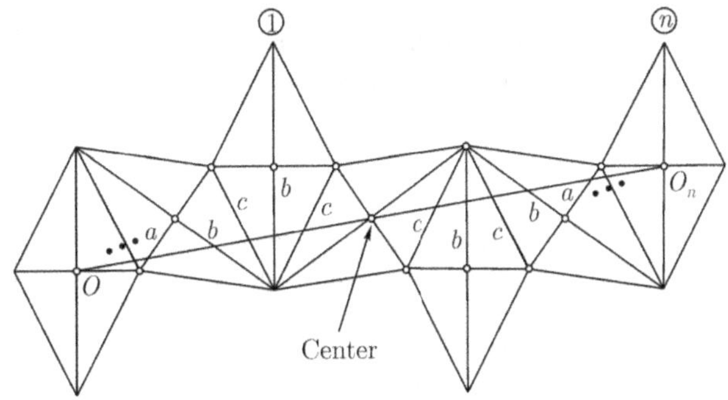

Fig. 3.40

of the triangle by a, the longer one by b, and the hypotenuse by c. Thus the sides of the rhombi are denoted by c and the halves of diagonals by a and b. Let us find the alternating sum of letters for mirror trajectories of the n-th type. By Proposition 3.6 the length of the period is even, so the sum is taken over one period. The unfolding of the trajectory is obtained from the line OO_n by a translation (see Fig. 3.40). Without taking into account the points where it passes through centers of rhombuses, we find the code word for the line OO_n. Since the line is always symmetric with respect to the center of one of the rhombuses, the code word has the form $X\overline{X}$, where X is a word and $\overline{X}$ is X read backwards. Now let us move the line OO_n slightly to the left so that it becomes a real unfolding of a trajectory. If n is even ($n = 2k$), the middle rhombus is inclined to the left. Therefore the code word will become $W = abXba\overline{X}$. By Proposition 3.6 the length of the word W is $10n + 6$. Therefore the length of X equals $\frac{10n+6-4}{2} = 10k + 1$, which is odd. Thus the alternating sum of letters of W is equal to

$$a - b + X - b + a - \overline{X} = 2a - 2b,$$

where X and $\overline{X}$ denote the alternating sum of the letters composing them. For an odd n, i.e., $n = 2k + 1$, the middle rhombus is inclined to the right, so after the shift of the line OO_n, the word becomes $W = abXab\overline{X}$, where the length of X which is equal to $10k + 6$, is even. The alternating sum equals

$$a - b + X + a - b + X = 2a - 2b.$$

If the length of X is even, then $X - \overline{X} = 0$, while if the length is odd, then $X + \overline{X} = 0$. Therefore the alternating sum is always equal to $2a - 2b$. Similarly, one can prove that if the unfolding is obtained from the line OO_n by a shift to the right, the alternating sum equals $2b - 2a$. The sum is different from 0, therefore mirror trajectories are unstable. This completes the part A.

B. We will use the following theorem.

Theorem 3.9. [39] *If a perpendicular periodic trajectory in a polygon is stable then the number of its links is a multiple of 4.*

It will be shown that the number of links in a perpendicular trajectory in a right triangle is never divisible by 4, which will imply its instability. It suffices to consider the case where the angle α is incommensurable with π, because if there existed a stable periodic trajectory for an α commensurable with π, then one would be able, by a small transformation of the triangle, to obtain such a trajectory for an incommensurable α as well.

In [26], it was shown that every perpendicular trajectory has the following form: Consider the unfolding of the trajectory such that the trajectory is perpendicular to a side or diagonal of the first rhombus. Let us assign a number to each of its rhombi: the initial rhombus has the number 0; a rhombus turned by an angle $n \cdot 2\alpha$ with respect to the initial one, has the number n.[5] Note that since α is incommensurable with π, the position of every rhombus determines its number unambiguously, and the numbers of two consecutive rhombi differ by 1. The particle launched from rhombus number 0 goes through rhombi with positive numbers until it meets another rhombus with number 0. Instead of continuing the unfolding in the same direction, we now imagine that the particle is reflected by the segment that is perpendicular to the trajectory. Then it returns to the initial rhombus following its own track.

Consider the first half of the track, before the reflection. The particle goes forward or backward depending on whether the numbers of rhombuses it goes through increase or decrease. Since the particle starts its motion going forward, i.e., from rhombus 0 to rhombus 1 and finishes it going backward, i.e., from rhombus 1 to rhombus 0, the total number of places where it changes direction is odd. Now note that, if a particle goes through a rhombus without changing direction, then this rhombus contains two links of the trajectory (see Fig. 3.41, (a)).

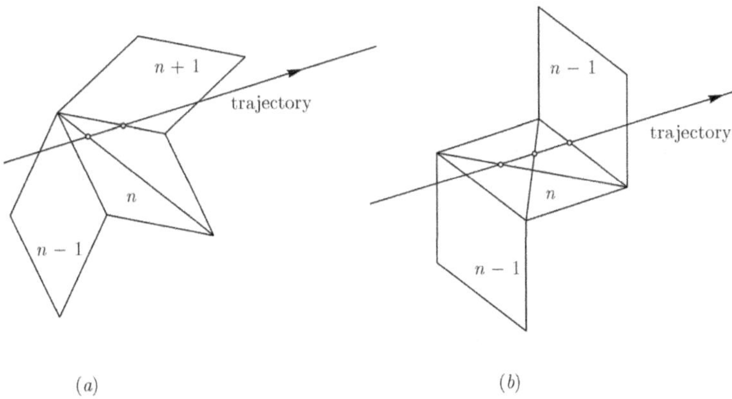

Fig. 3.41

In contrast, when the particle changes direction inside a rhombus, the rhombus contains three of its links (Fig. 3.41, (b)).

Since the number of rhombi where it changes direction is odd, the total number of links in the first half of the track is also odd. The whole period contains twice as many sides as its first half, so their total number cannot be divisible by 4. Thus the instability is proved.

---

[5]The same number can be assigned to several rhombuses.

C. By Proposition 3.5, part i), the S-trajectories are perpendicular. Therefore it follows that the S-trajectories are unstable. Using the same method as in Part A one can show that the alternating sum for S-trajectories is equal to $\pm(2a - 2b)$. $\qquad\square$

In the end of this section we give (without proofs) the following results of [58], [59].

Theorem 3.10. [58] *Right triangles have no stable periodic billiard trajectory.*

Theorem 3.11. [59] *An isosceles triangle has a stable periodic billiard trajectory if and only if its base angles are not $\frac{\pi}{2n}$ where n is a power of 2.*

Definition 3.4. An isosceles triangle with base angle $\frac{\pi}{2n}$ is called the Veech point V_n for all $n = 2, 3, \ldots$.

Theorem 3.12. *The isosceles triangle V_n has a stable periodic billiard trajectory if and only if n is a power of 2.*

For $n = 2$, V_2 is a right triangle, so we know it has no periodic billiard trajectory.

Exercise 27. [30] Consider the altitude construction of a 3-periodic path in a triangle.

(a) Sketch a trajectory that is parallel to the one in the construction but that starts a small distance from the base point of the altitude. Show that this trajectory is also periodic, and find its period.

(b) Explain why the altitude construction of a 3-periodic path only works for acute triangles.

(c) Does every obtuse triangle have a periodic billiard path? This question is open, and is the subject of current research. The answer is known to be "yes" for many cases. Find an example of a periodic path in an obtuse triangle.

3.4 Elliptical billiard tables

3.4.1 *Reflection law of ellipse*

We will consider a billiard table that takes the shape of an ellipse. This section will be mainly based on [86].

The bouncing law is linked to the reflection law in optics. First, we recall some facts about ellipses.[6]

Definition 3.5. Consider two given points F_1 and F_2 in a plane, an ellipse is the locus of all points P in that plane for which the sum of the distances to F_1 and F_2 equals a given value $2a$, i.e.

$$|PF_1| + |PF_2| = 2a.$$

The given points F_1 and F_2 are called the focal points of the ellipse, PF_1 and PF_2 are called the focal radii of the point P.

Note that for all points outside the ellipse the sum of the distances to the focal points is larger than $2a$, for all points inside the ellipse it is smaller than $2a$.

The midpoint C of the line segment joining the focal points is called the center of the ellipse. The line through the focal points is called the major axis, and the line perpendicular to it through the center is called the minor axis. It contains the vertices V_1, V_2, which have distance a to the center. The distance c of the focal point to the center is called the focal distance or linear eccentricity. The quotient $\frac{c}{a}$ is the eccentricity e (see Fig. 3.42).

If Cartesian coordinates are introduced such that the origin is the center of the ellipse and the x-axis is the major axis and the foci are the points $F_1 = (c, 0)$, $F_2 = (-c, 0)$, and a and b are the semi-major and semi-minor axes respectively. Then we have the relation $b^2 = a^2 - c^2$ using this we obtain the equation of the ellipse:

$$\frac{x^2}{a^2} + \frac{y^2}{b^2} = 1. \tag{3.4}$$

The following property is very useful in the further treatment of billiards in the ellipse (see Fig. 3.43):

Lemma 3.1. *For any point P on the ellipse, the bisectors b and b' of FP and $F'P$ are the tangent and normal of the ellipse at P.*

Proof. Consider the orthogonal line from F to b which meets b in Q and $F'P$ in F''.

For any point R (different from P) on b we have: $RF = RF''$. Moreover $RF' + RF'' > F'F'' = F'P + PF = 2a$.

From the definition of the ellipse, it follows that R is outside of the ellipse, and so are all points on b different from P. Thus, the bisectors b and b' are the tangent and normal of the ellipse at P. $\qquad\square$

---

[6]http://cage.ugent.be/ hs/billiards/billiards.html

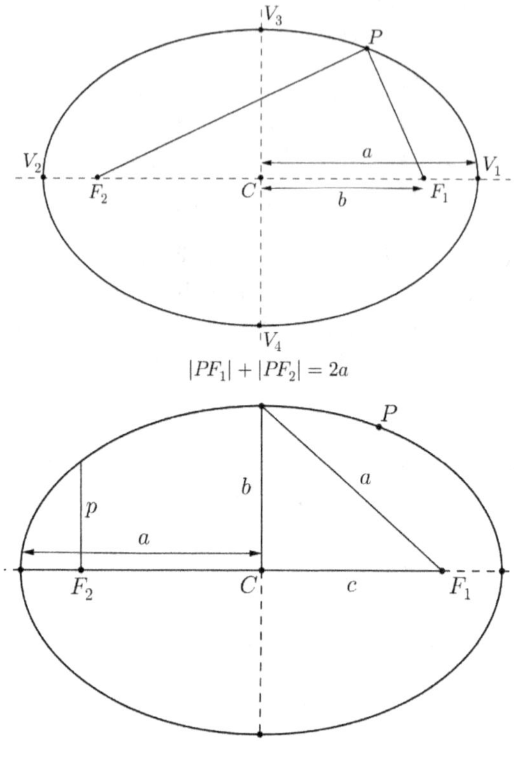

Fig. 3.42

Remark 3.6. An inscribed regular polygon in an ellipse is a regular polygon whose vertices all lie on a single ellipse. Note that in an ellipse it is possible to inscribe only a regular triangle and a square, because the inscribed regular polygon should be inscribed at the same time in the circle which is crossed with an ellipse in no more than four points – the polygon's vertices.

Exercise 28. [30] (a) Consider the billiard trajectory $A_0 A_1 A_2$ in the larger ellipse E shown in the Fig. 3.44. Prove that $\angle A_0 A_1 F_1 = \angle A_2 A_1 F_2$.

(b) Reflect F_1 across $\overline{A_0 A_1}$ to create F_1' and reflect F_2 across $\overline{A_1 A_2}$ to create F_2'. Prove that

$$\angle A_0 A_1 F_1' = \angle A_0 A_1 F_1, \quad \text{and} \quad \angle A_2 A_1 F_2' = \angle A_2 A_1 F_2.$$

(c) Show that $\Delta F_1' A_1 F_2$ and $\Delta F_1 A_1 F_2'$ are congruent.

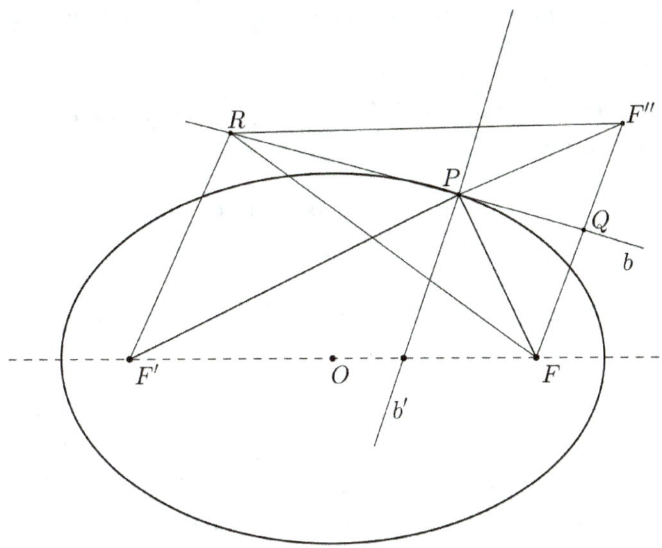

Fig. 3.43

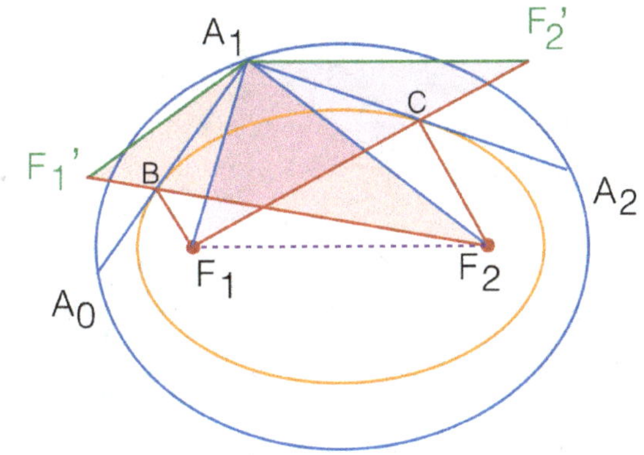

Fig. 3.44

(d) Mark the intersection of $\overline{F_1'F_2}$ with $\overline{A_0A_1}$ as B, and the intersection of $\overline{F_1F_2'}$ with $\overline{A_1A_2}$ as the ellipse C. Show that the string length $|\overline{F_1B}| + |\overline{BF_2}|$ is the same as the string length $|\overline{F_1C}| + |\overline{CF_2}|$.

The following property of the ellipse is similar to the billiard reflection law.

Theorem 3.13. *The focal radii F_1A and F_2A form equal angles with the norm in A (see Fig. 3.45).*

Proof. Consider rays of the segments F_1A and F_2A outside the ellipse, both in the direction from the focal points to point A. On the ray F_1A take a point F_2' such that the length of the segment $F_2'A$ is equal to F_2A, i.e., $|F_2'A| = |F_2A|$. Similarly, on the ray F_2A, take a point F_1' such that $|F_1'A| = |F_1A|$. It is easy to see that the triangles $\triangle AF_1F_1'$ and $\triangle AF_2F_2'$ are isosceles and that they are similar to each other. Consider now a line m which passes through the point A and bisects $\angle F_1AF_2$. Let G be the intersection point between the line m and the segment F_1F_2. It is clear that $\angle F_1AF_2 = 2\angle F_2'F_2A$ and consequently, $\angle GAF_2 = \angle F_2'F_2A$. Hence line m and segment F_2F_2' are parallel. Notice that line l bisects $\angle F_2AF_2'$. Thus the segment F_1A and F_2A make the same angle with line l. $\qquad\square$

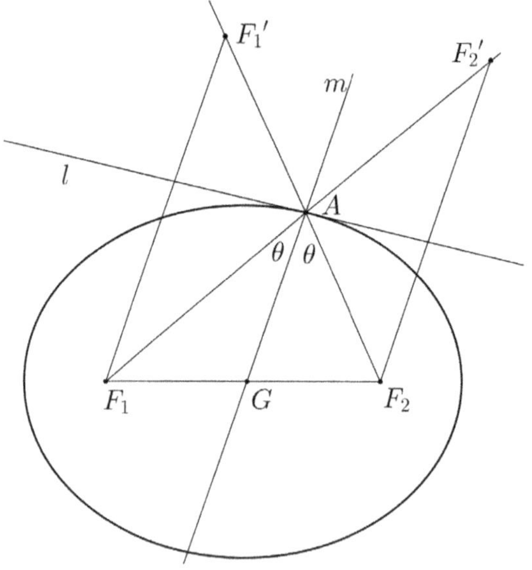

Fig. 3.45

Consider a planar billiard table D with boundary Γ.

Definition 3.6. A caustic is a curve inside a planar billiard table such that if a segment of a billiard trajectory is tangent to this curve, then so is each reflected segment.

Assume that caustics are smooth and convex.

Note that billiards in a circle has a family of caustics, consisting of concentric circles.

Let Γ be a billiard curve and γ a caustic.

Erase the boundary Γ of the billiard table and only the caustic γ remains. Is it possible to recover Γ back from γ?

The answer is yes! This is given by the following string construction: wrap a closed non-stretchable string around γ, pull it tight at a point and move this point around γ to obtain a curve Γ.

Now we discuss three cases of a billiard trajectory considered on an elliptic table. In two of these cases there are caustic curves (one is an ellipse the second one is a hyperbola).

3.4.2 *First case: Ball passes along a focus*

Consider billiards on elliptic table, with the ball positioned at P_1 and shot in the direction of the focus F_1. If the ball hits the border of the table at P_2 it will be bounced according to the law of reflection: the angle between the incoming path and the normal at P_2 equals the angle between this normal and the outgoing path. As a consequence of the property last mentioned the ball will bounce back along P_2F_2. After a new hit, the ball will bounce through the other focus F_1.

The following theorem states that after a rather small number of bounces the ball travels along the major axis.

Theorem 3.14. *Let Q be an elliptical billiard table in $\mathbb{R}^2$. Let $\{a_n\}$ be the sequence of points on the boundary ∂Q where the billiard contacts the boundary of Q. Moreover, for every non-negative integer n the segment $\overline{a_n a_{n+1}}$ crosses the two focal points. Then the trajectory of the billiard converges to the major axis of the ellipse (see Fig. 3.46). That is, for any ε neighborhood of the major axis there is an n_0 such that for any $n > n_0$ the segment $\overline{a_n a_{n+1}}$ lies in the neighborhood.*

Proof. Consider two lines l and m parallel to the minor axis (see Fig. 3.47), each passing through the focus of the ellipse. Using these lines, divide the boundary of the elliptical billiard table into three sections:

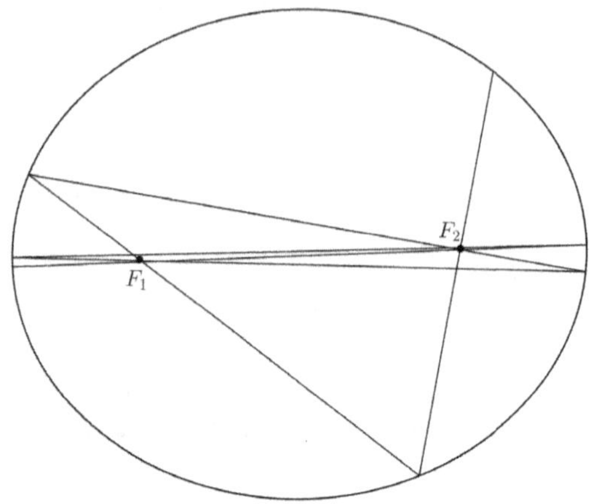

Fig. 3.46

- the closed boundary left to the line l, denoted by ∂Q_1,
- the boundary between the two lines, denoted by ∂Q_2,
- the boundary right to the line m, denoted by ∂Q_3.

The proof of the theorem is divided in two steps:

Step 1. First we show that the set of trajectories W_1 which starts from ∂Q_1 and passes through F_2 converges to the major axis.

It is easy to see that the angle of incidence is 0 if and only if the trajectory is along the major axis of the ellipse. Therefore we shall show that the sequence of angle of incidence of the set W_1, denoted by $\{\theta_{n_1}\}$, converges to 0. We prove that the sequence $\{\theta_{n_1}\}$ is monotonically decreasing. Let a_k, a_{k+1}, and a_{k+2} be three consecutive elements of the sequence $\{a_n\}$. Let $a_k \in \partial Q_1$ and line l be perpendicular to the segment $\overline{a_k a_{k+1}}$ which passes through point F_1. Let point B be the intersection between the line l and the segment $\overline{a_k a_{k+1}}$. Denote the segments as follows

$$x = \overline{a_k F_1}, \quad y = \overline{a_{k+1} F_1}, \quad z = \overline{F_1 B}.$$

Denote (see Fig. 3.47) one point of the intersections between the ellipse and line l by P and denote one point of the intersections between the ellipse and line m by Q. Observe that the segment $\overline{PF_1}$ is the longest among the set of segments $\overline{SF_1}$ for $S \in \partial Q_1$. The segment $\overline{QF_1}$ is the shortest among the set of segments $\overline{TF_1}$ for $T \in \partial Q_3$. Denote the length of the minor axis as $2a$, the major axis as $2b$, and

$$\alpha = \overline{QF_1}, \quad \beta = \overline{PF_1}, \quad 2f = \overline{F_1 F_2}.$$

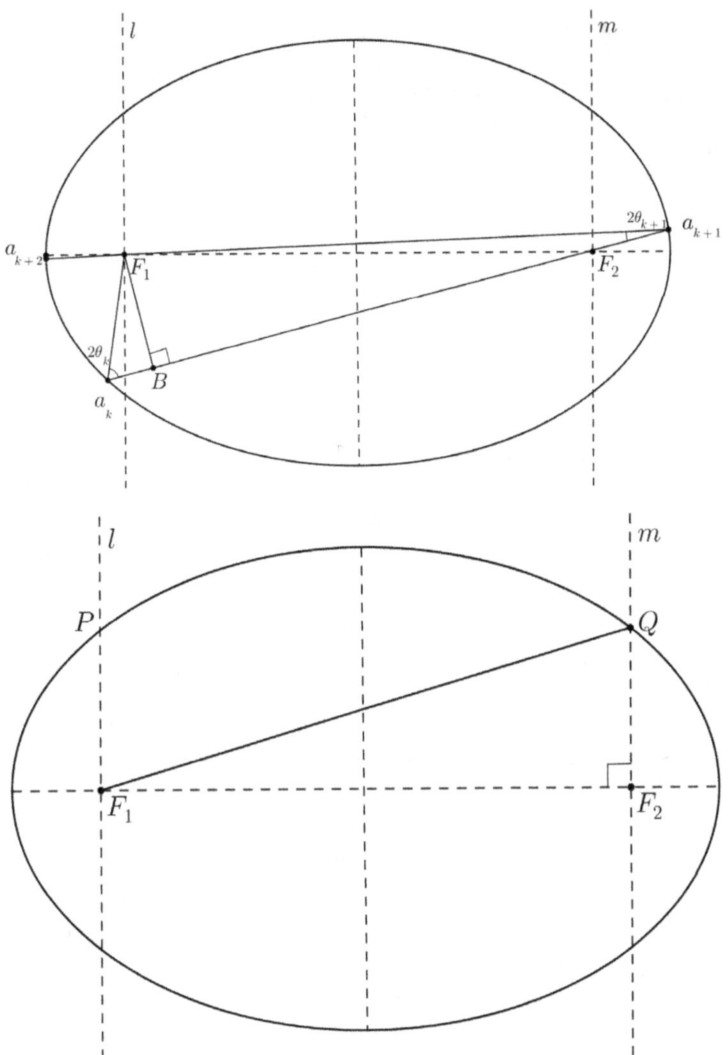

Fig. 3.47

Since $\overline{PF_1} = \overline{QF_2}$ and $\triangle QF_1F_2$ is a right triangle, we have

$$\alpha + \beta = 2b,$$
$$\alpha^2 - \beta^2 = 4f^2 = 4(b^2 - a^2).$$

Solving the system of equations we get

$$\alpha = 2b - \frac{a^2}{b}, \quad \beta = \frac{a^2}{b}.$$

We have

$$\alpha - \beta = 2b - 2\frac{a^2}{b} = \frac{2}{b}(b^2 - a^2) = \frac{2(a+b)}{b}(b-a) \geq 0, \quad \text{since} \ \ b \geq a.$$

The equality $\alpha = \beta$ holds when the ellipse is a circle (i.e. $a = b$). Thus for arbitrary points $S \in \partial Q_1$ and $T \in \partial Q_3$ the following holds.

$$x = \overline{SF_1} \leq \overline{PF_1} \leq \overline{QF_1} \leq \overline{TF_1} = y.$$

Hence $x \leq y$. This implies that

$$\sin(2\theta_k) = \frac{z}{x} \geq \frac{z}{y} = \sin(2\theta_{k+1}).$$

Since $\sin(\theta)$ is an increasing function on the interval $[-\pi/2, \pi/2]$, we have $\theta_k \geq \theta_{k+1}$ for any k, i.e., the sequence $\{\theta_k\}$ is monotonically decreasing. The sequence is bounded by the interval $[-\pi/2, \pi/2]$, therefore the sequence converges, which implies the billiard trajectory also converges to a segment in the ellipse.

Now we show that the sequence converges to 0. Assume that the sequence converges to a non-zero value θ. Notice that the billiard trajectory has to pass through either of the two focus points. Then the trajectory has to oscillate between two parallel segments, each passing through one of the focus. This is clearly a contradiction since the billiard trajectory has to converge. Even if we assume that the trajectory converges to one of the two segments, the consecutive trajectory has angle of incidence less than θ, another contradiction. Thus the billiard trajectories in the set W_1 converge to the major axis of the ellipse.

Step 2. We now show that the first step is equivalent to proving the theorem. Except for the case in which the billiard trajectory is on the major axis or the minor axis of the ellipse, at some point the set of trajectories has to intersect with ∂Q_2. Assume that the set of billiard trajectories W_2 starts from a point on ∂Q_1 but crosses the focus F_1. Note that observing W_2 in time is equivalent to observing W_1 backwards in time. Since the sequence $\{\theta_{n_1}\}$ monotonically decreases, the sequence $\{\theta_{n_2}\}$, which is the sequence of angles of incidence in W_2, monotonically increases as long as W_2 intersects at the interiors of ∂Q_1 and ∂Q_3.

Denote by G (see Fig. 3.48) one of the points of the intersections between line m and ∂Q_3 and let point H be the intersection between the ray GF_1

and ∂Q. Define $\angle F_1 H F_2$ as the critical angle of incidence, denoted as $2\theta_{\rm c}$. It is clear that there exists a finite number m such that the angle of incidence θ_{m_2} at point a_m of the set W_2 eventually equals or exceeds $\theta_{\rm c}$. Without loss of generality assume a_m is in ∂Q_1.

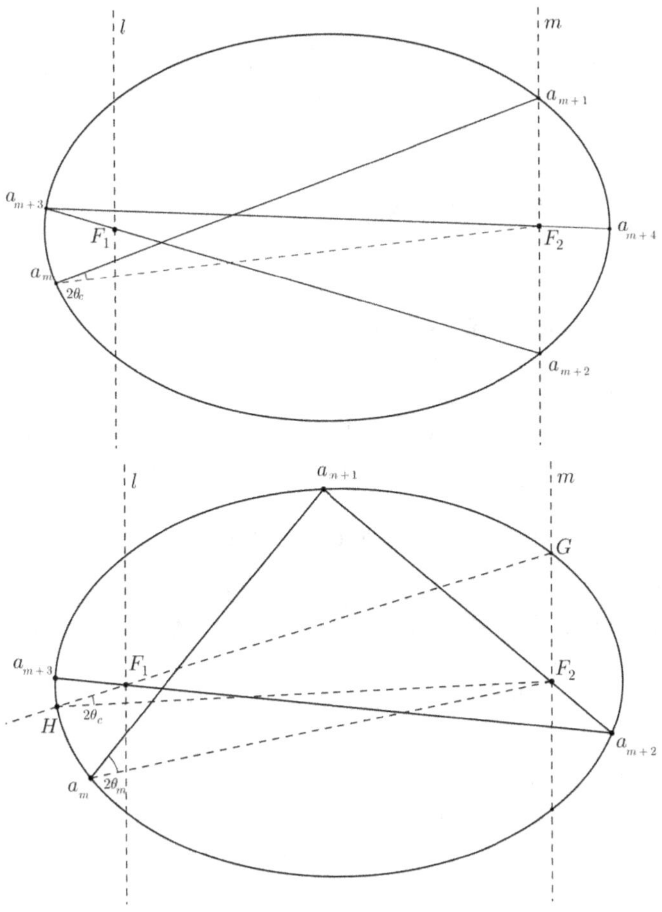

Fig. 3.48

If $\theta_{m_2} = \theta_{\rm c}$ or a_m is identical to point H and a_{m+1} is identical to point G, then a_{m+3} is on ∂Q_1 while the trajectory $\overline{a_{m+3}a_{m+4}}$ passes through F_2. Consequently, W_2 follows the trajectory W_1 after $m+3$ iterations of billiard mapping.

For the case $\theta_{m_2} > \theta_c$ we have a_{m+1} is on ∂Q_2. Consequently, the trajectory $\overline{a_{m+2}a_{m+3}}$ is such that ether $a_{m+2} \in \partial Q_1$ and $\overline{a_{m+2}a_{m+3}}$ crosses through F_2 or $a_{m+2} \in \partial Q_3$ and $\overline{a_{m+2}a_{m+3}}$ crosses through F_1. Therefore, every trajectory, after $m+2$ iterations of billiard mapping, in elliptical billiard table is equivalent to the trajectory W_1. This completes the proof.

$\square$

If the first shot does not pass over a focus none of all subsequent reflected paths will pass over a focus. Still, in this case the mathematical elliptical billiard game leads to interesting patterns.

3.4.3 *Second case: The first shot passes between the foci of the ellipse*

It follows from the mentioned property of the focal radii in relation to the normal or tangent of the reflection point that all subsequent reflected segments will pass between the foci of the ellipse. If we consider a large number of reflected segments there seems to arise a new curve. Below we will see that this curve, known as the evolute of the reflected segments, is a hyperbola and moreover the foci of this hyperbola are the same as those of the original ellipse.

The definition of a hyperbola resembles the definition of an ellipse:

Definition 3.7. Given two points F_1 and F_2 in a plane, a hyperbola is the locus of all points P in that plane for which the difference of the distances to F_1 and F_2 equals a given value $2a$. The given points F_1 and F_2 are called the foci of the hyperbola and PF_1 and PF_2 are called the focal radii of the point P.

Theorem 3.15. *Let Q be a hyperbola with foci F_1 and F_2. Let A be a point on the boundary of the hyperbola. Draw line l which is tangent to the hyperbola at point A. Then the segments F_1A and F_2A make the same angle with line l (see Fig. 3.49).*

Proof. The proof is similar to the proof of Theorem 3.13. $\square$

The following theorem explains the case where the billiard trajectory crosses the line segment between the two foci of the ellipse.

Theorem 3.16. *Let Q be an elliptical billiard table in $\mathbb{R}^2$. Let $\{a_n\}$ be the sequence of points on ∂Q where the billiard contacts the boundary of Q. Moreover, for every non-negative integer n the segment $\overline{a_n a_{n+1}}$ intersects*

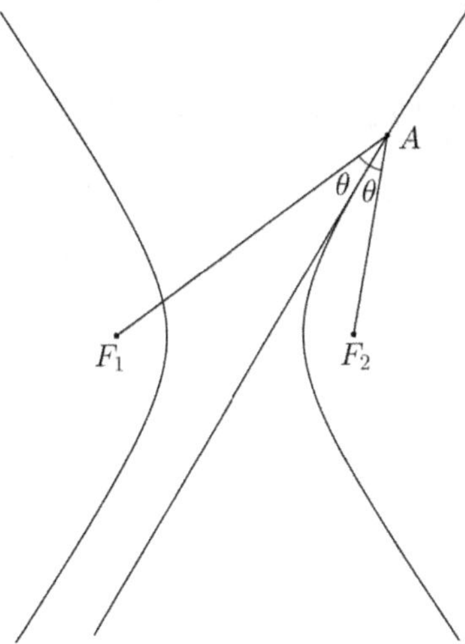

Fig. 3.49

with the line segment between the two foci of the ellipse. Then the trajectory of the billiard is tangent to the hyperbola which shares the same foci with the ellipse Q. Namely, the trajectory has a caustic which is a hyperbola sharing the same foci with the ellipse.

Proof. Consider three consecutive elements of the sequence $\{a_n\}$,

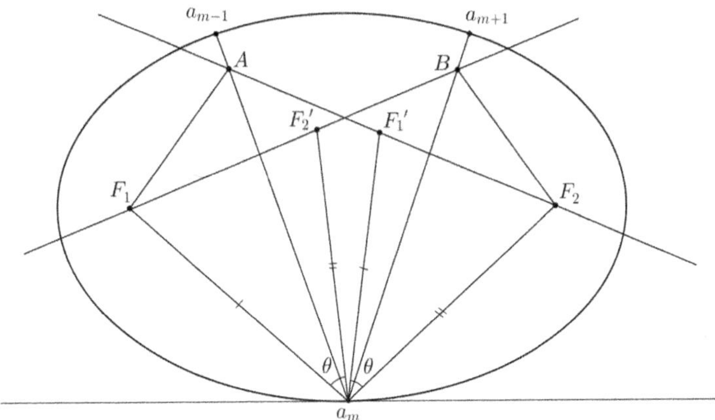

Fig. 3.50

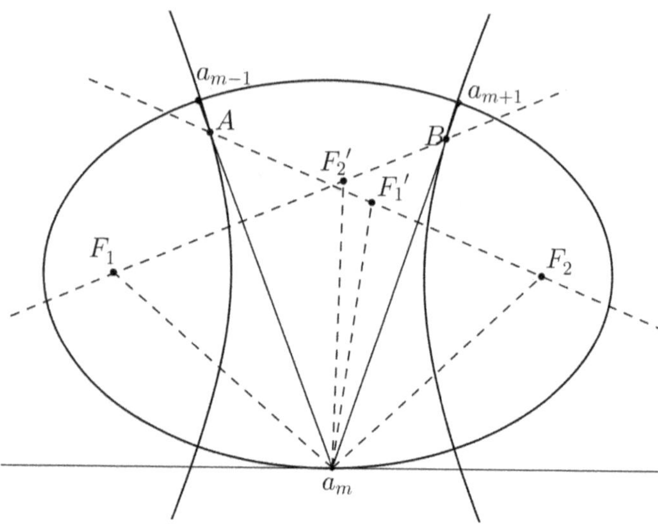

Fig. 3.51

namely a_{m-1}, a_m, and a_{m+1} (see Figs. 3.50 and 3.51). Denote by F_1' the reflection point of F_1 with respect to the segment $\overline{a_{m-1}a_m}$ and let point F_2' be the reflection point of F_2 with respect to the segment $\overline{a_m a_{m+1}}$. Let point A be the intersection of the line $F_1'F_2$ and $\overline{a_{m-1}a_m}$ and let point B be the intersection of the line $F_2'F_1$ and $\overline{a_m a_{m+1}}$. Since F_1' is a reflection point of F_1, we have $\angle a_{m-1}AF_1' = \angle a_{m-1}AF_1$, which implies $\angle a_{m-1}AF_1 = \angle a_m AF_2$. By Theorem 3.15, A is the point where the segment $\overline{a_{m-1}a_m}$ is a tangent to the hyperbola Q_1 with the foci F_1 and F_2. Similarly, B is the point where the segment $\overline{a_m a_{m+1}}$ is tangent to the hyperbola Q_2 with the foci F_1 and F_2. Now we show $Q_1 = Q_2$ by showing

$$|\overline{F_1 A} - \overline{F_2 A}| = |\overline{F_1 B} - \overline{F_2 B}|,$$

this is equivalent to $\overline{F_1'F_2} = \overline{F_1 F_2'}$. Using Theorem 3.13 we get $\angle F_1' a_m F_2 = \angle F_1 a_m F_2'$, which implies $\triangle F_1' a_m F_2 \equiv \triangle F_1 a_m F_2'$. Consequently, $\overline{F_1'F_2} = \overline{F_1 F_2'}$. $\qquad\square$

3.4.4 *Third case: The first shot does not pass between the foci of the ellipse*

It follows from the mentioned property of the focal radii in relation with the normal or tangent at the reflection point that none of all subsequent reflected segments will pass between the foci of the ellipse. If we consider a large number of reflected segments there arises a new ellipse. The following theorem states that this curve, known as the evolute of the reflected segments is indeed an ellipse. Moreover, the foci of this ellipse are the same as those of the original ellipse.

Theorem 3.17. *Let Q be an elliptical billiard table in $\mathbb{R}^2$. Let $\{a_n\}$ be the sequence of points on ∂Q where the billiard contacts the boundary of Q. Moreover, for every non-negative integer n, the segment $\overline{a_n a_{n+1}}$ does not intersect with the line segment between the two foci of the ellipse. Then every trajectory of the billiard is tangent to the ellipse which shares the same foci with the ellipse Q. In other words, the trajectory has a caustic which is an ellipse confocal to the elliptical billiard table.*

Proof. The proof is similar (using Fig. 3.52) to the proof of Theorem 3.16. $\qquad\square$

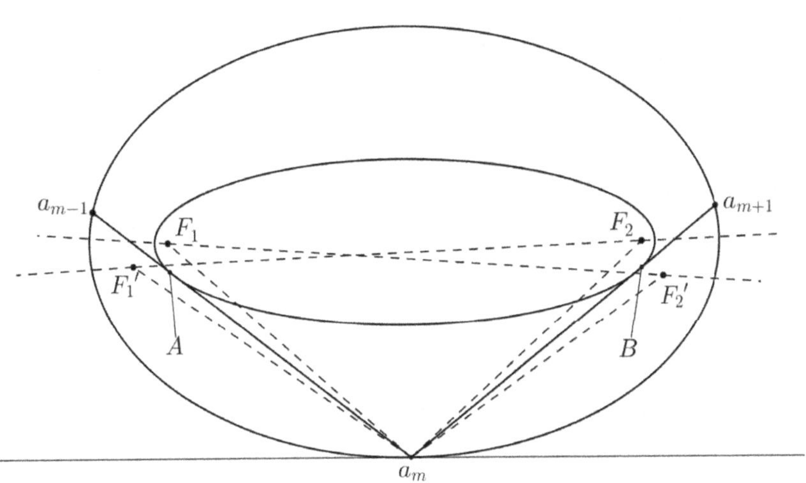

Fig. 3.52

Thus we have the situations shown in Fig. 3.53

Some exceptions: periodical paths. One easily can see that there exist particular situations. In some cases the paths are periodical and do not lead to a new ellipse inside the given one. We give two examples: a quadrangular path and a hexagonal path.

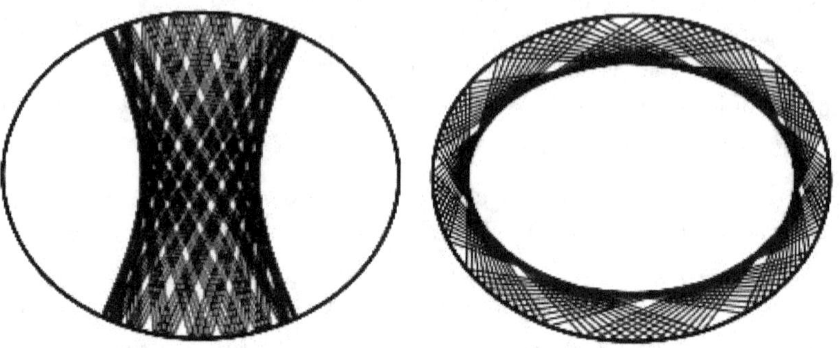

Fig. 3.53 The trajectory has a caustic which is a hyperbola (left) and an ellipse (right) confocal to the elliptical billiard table.

A circle can be considered as a special ellipse: the foci F_1 and F_2 coincide. If a path is non-periodic the evolute of the reflected shots is a new circle with the same center. The existence of periodical paths is easier to see, which was discussed in Chapter 2.

If we push the ball along the minor axis of the ellipse it will continuously bounce along the same axis.

Exercise 29. Show that the caustic of a circle is another circle, and find its radius as a function of the angle α between the billiard trajectory and the (tangent line to the) circle.

3.5 Birkhoff theorems

3.5.1 *Caustics and mirror equation*

In this subsection following [110] we give some geometry of billiard caustics.

Consider a planar billiard table D whose boundary is a smooth closed curve Γ. Let M be the space of unit tangent vectors (x, v) (where v is the velocity vector) whose foot points x are on Γ and have inward directions.

A vector (x, v) is an initial position of the billiard ball. The ball moves freely and hits Γ at point x_1; let v_1 be the velocity vector reflected off the boundary.

The billiard ball map $T : M \to M$ maps (x, v) to (x_1, v_1). In general, if D is not convex, then T is not continuous: this is due to the existence of billiard trajectories touching the boundary from inside.

If we give a parameterization of Γ by arc length t and let α be the angle between v and the positive tangent line of Γ, then (t, α) are coordinates on M. In particular, M is a cylinder.

Moreover, if Γ is a strictly convex closed billiard curve, then the space M of the billiard ball map T consists of oriented lines that intersect Γ. It is a subset of the space N of all oriented lines in the plane.

An invariant circle of the billiard ball map T is a simple closed T-invariant curve $\delta \subset M$ that makes one turn around the phase cylinder.

For example, if Γ is a circle, then M is foliated by invariant circles; and if Γ is an ellipse, then part of M, containing the boundary, is foliated by invariant circles.

Assuming that an invariant circle δ is a smooth curve, one can consider δ as a smooth one-parameter family of oriented lines intersecting the billiard table.

Let $\delta \subset M$ be an invariant circle of the billiard ball map inside Γ and γ be the respective caustic.

We are going to answer the question: can γ have points outside of Γ? To answer this question, one needs the following

Theorem 3.18. (Birkhoff) *Under coordinates* (t, α) *in* M, *the curve* δ *is the graph* $\alpha = f(t)$ *of a continuous function* f.

Proof. To prove this theorem, we use a class of area preserving twist maps of the cylinder. The twist condition for a map $T : (t, \alpha) \to (t_1, \alpha_1)$ means that $\partial t_1 / \partial \alpha > 0$. This condition holds for the billiard ball map in a convex billiard (see Fig. 3.54). See, e.g., [65] for the theory of twist maps and for a proof of the Birkhoff theorem. $\qquad\square$

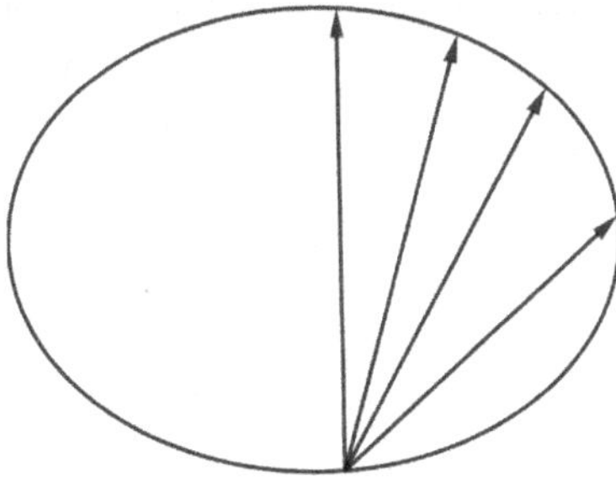

Fig. 3.54　The twist condition for a convex billiard table: $t =$const.

Lemma 3.2. *Let γ be the caustic corresponding to an invariant circle δ of the billiard ball map inside a convex curve Γ. Then γ lies inside Γ.*

Proof. By Theorem 3.18 we have that the curve δ is a graph of $\alpha = f(t)$ and therefore, the map T, restricted to δ, can be written as

$$T(t, f(t)) = (g(t), f(g(t))),$$

where g is a monotonically increasing function. Let $t_1 = t+\varepsilon$ be a point close to t. Then the straight lines $(\Gamma(t), \Gamma(g(t)))$ and $(\Gamma(t_1), \Gamma(g(t_1)))$ intersect in the interior of Γ (see Fig. 3.55). Since ε is small, taking $\varepsilon \to 0$ completes the proof.

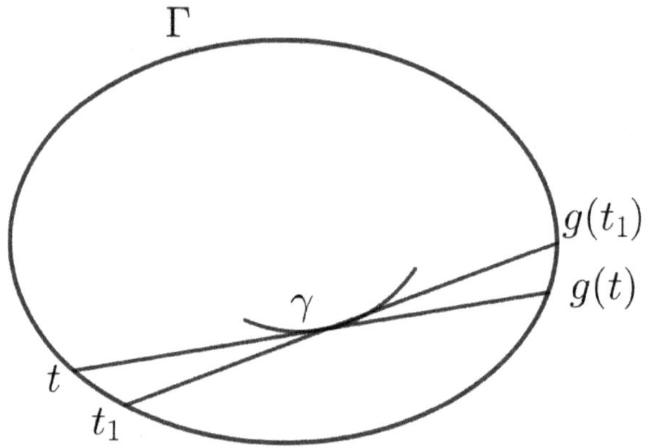

Fig. 3.55

$\square$

This lemma fails for the confocal hyperbolic caustic of the billiard inside an ellipse because the respective invariant curves in the phase cylinder are contractible and do not make a turn around the cylinder.

In geometrical optics very useful is an equation called the mirror equation. Here again following [110] we give this formula.

Let Γ be the boundary of a billiard table, which is the reflection cure. Assume that an infinitesimal beam of light with center A reflects to a beam with center B (Fig. 3.56). Let X be the reflection point. Denote the equal angles made by AX and BX with Γ by α. Co-orient Γ by the unit normal n that has the inward direction, and let $\mathcal{K}$ be the curvature[7] of Γ at point

---

[7]Recall that for a plane curve Γ given by parametric equations $x = x(t)$ and $y = y(t)$, the curvature $\mathcal{K}$ is defined by $\mathcal{K} = \frac{x'y''-y'x''}{(x'^2+y'^2)^{3/2}}$.

X. Note that $\mathcal{K}$ is positive if the billiard table is convex outward, and it is negative otherwise.

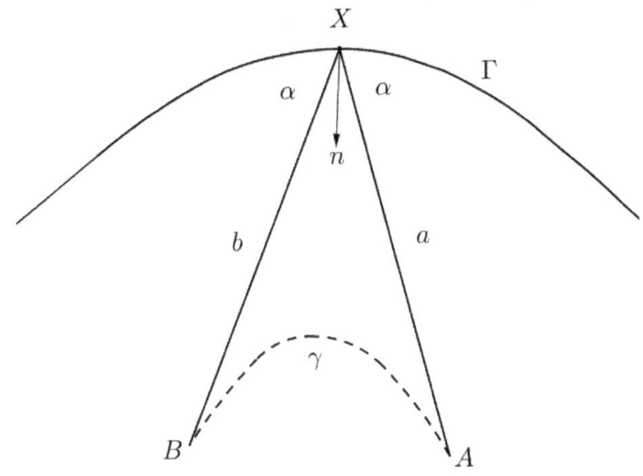

Fig. 3.56

Denote by a and b the signed distances from points A and B to X respectively. By convention, $a > 0$ if the incoming beam focuses before the reflection, and $b > 0$ if the reflected beam focuses after the reflection.

Theorem 3.19. *The following relation holds*

$$\frac{1}{a} + \frac{1}{b} = \frac{2\mathcal{K}}{\sin \alpha}. \tag{3.5}$$

Proof. Let Γ be parameterized by arc length parameter t so that $X = \Gamma(0)$. Define the function

$$f(t) = |\Gamma(t) - A| + |\Gamma(t) - B|.$$

Since the ray AX reflects to XB, we have: $f'(0) = 0$. Since infinitesimally close rays from A also reflect to rays through B, one also knows that $f''(0) = 0$. Let us express these conditions in terms of the given data. Noting that $a = |\Gamma(t) - A| = \sqrt{(\Gamma(t) - A)^2}$, we obtain

$$a' = |\Gamma(t) - A|' = \frac{(\Gamma(t) - A)\Gamma'(t)}{a} = \cos \alpha,$$

and similarly $|\Gamma(t) - A|' = -\cos \alpha$. Computing the second derivation we obtain

$$|\Gamma(t) - A|'' = \left(\frac{(\Gamma(t) - A)\Gamma'(t)}{a} \right)'$$

$$= \frac{[(\Gamma(t) - A)\Gamma'(t)]'a - (\Gamma(t) - A)\Gamma'(t)a'}{a^2}$$

$$= \frac{\Gamma'(t)\Gamma'(t)}{a} + \frac{(\Gamma(t) - A)\Gamma''(t)}{a} - \frac{((\Gamma(t) - A)\Gamma'(t))^2}{a^3}$$

$$= \frac{1}{a} - \mathcal{K}\sin\alpha - \frac{\cos^2\alpha}{a} = \frac{\sin^2\alpha}{a} - \mathcal{K}\sin\alpha.$$

Similarly,

$$|\Gamma(t) - B|'' = \frac{\sin^2\alpha}{b} - \mathcal{K}\sin\alpha.$$

Since $f''(0) = 0$, we have

$$\frac{\sin^2\alpha}{a} + \frac{\sin^2\alpha}{b} - 2\mathcal{K}\sin\alpha = 0.$$

From the last equality we get (3.5). $\qquad\qquad\qquad\qquad\qquad\qquad$ $\square$

Example 3.3. If Γ is a straight line, then $\mathcal{K} = 0$ and $b = -a$, then the focusing point of the reflected beam is behind the mirror.

By the mirror equation we get the following property of caustics: a point of a caustic is the focus of an infinitesimal beam that focuses, after reflection, at another point of this caustic (see Fig. 3.56). This implies the following phenomenon discovered by J. Mather [79].

Proposition 3.7. *If the curvature of a convex smooth billiard curve vanishes at some point, then this billiard ball map has no invariant circles.*

Proof. Suppose that there is an invariant circle and let $\gamma \subset \Gamma$ be the respective caustic. Let X be a point of zero curvature, and XA and XB be tangent to γ from point X, making equal angles with Γ. The mirror equation (3.5) implies that $b = -a$, and therefore one of the points A or B lies outside the billiard table. $\qquad\qquad\qquad\qquad\qquad\qquad$ $\square$

As was shown in previous sections the billiards in ellipses are integrable: the billiard table is foliated by caustics, the confocal ellipses, and part of the trajectory consists of oriented lines tangent to these caustics. The billiard in a circle is even more regular: every point is an oriented line, tangent to a caustic.

The following theorem (see [110], [120]) shows how exceptional the situation of ellipses for other tables.

Theorem 3.20. *If almost every phase point of the billiard ball map in a strictly convex billiard table belongs to an invariant circle, then the billiard table is a disc.*

Proof.   Let (x, v) be a point of the trajectory and let

$$T(x, v) = (x', v'), \quad T^{-1}(x, v) = (x'', v'').$$

Denote the chord length $|xx'|$ by $f(x, v)$. The line xx'' is tangent to a caustic γ; denote by $a(x, v)$ the length of its segment from the tangency point to x, see Fig. 3.57. Let $\mathcal{K}(x)$ be the curvature of the billiard curve. Using the formula (3.5) we get

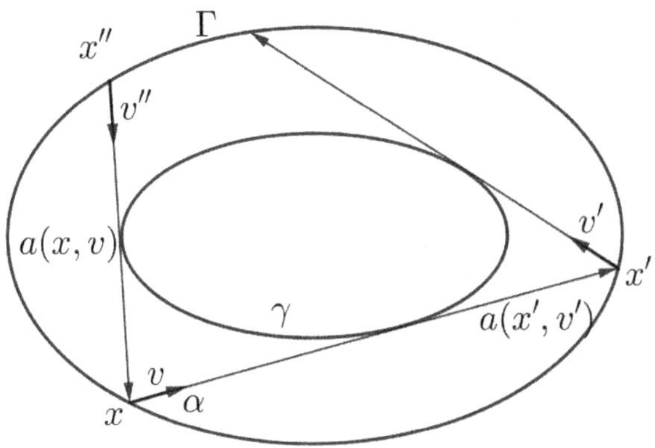

Fig. 3.57

$$\frac{1}{a(x, v)} + \frac{1}{f(x, v) - a(x', v')} = \frac{2\mathcal{K}(x)}{\sin \alpha},$$

$$\frac{4a(x, v)(f(x, v) - a(x', v'))}{a(x, v) + f(x, v) - a(x', v')} = \frac{2 \sin \alpha}{\mathcal{K}(x)}. \tag{3.6}$$

By applying the inequality between the harmonic and the arithmetic mean, i.e.,

$$AB \leq \left(\frac{A + B}{2}\right)^2,$$

we get

$$\text{LHS of } (3.6) \leq 4\frac{(f(x, v) + a(x, v) - a(x', v'))^2}{4(a(x, v) + f(x, v) - a(x', v'))} = f(x, v) + a(x, v) - a(x', v').$$

Let t be the arc length parameter on the billiard curve Γ and L its length. Integrate both sides of (3.6) over the phase space with respect to its T-invariant area form:

$$\int_M (f(x, v) + a(x, v) - a(T(x, v)))\omega = \int_M f(x, v)\omega = 2\pi A,$$

where A is the area of the table and $\omega = \sin\alpha d\alpha \cdot dt$. The integral of the other side of (3.6) equals

$$\int_0^L \int_0^\pi \frac{2\sin^2\alpha}{\mathcal{K}(t)} dt\, d\alpha = \pi \int_0^L \frac{1}{\mathcal{K}(t)} dt.$$

Recall the Cauchy-Schwartz inequality:

$$\int_0^L g^2(t)dt \int_0^L h^2(t)dt \geq \left(\int_0^L g(t)h(t)dt\right)^2.$$

Using this inequality we get

$$\int_0^L \frac{1}{\mathcal{K}(t)} dt \int_0^L \mathcal{K}(t)dt \geq L^2.$$

By the equality $\int_0^L \mathcal{K}(t)dt = 2\pi$, we get $2\pi A \geq L^2/2$. This is opposite to the isoperimetric inequality (3.7) (see below), hence it is actually an equality, and the curve Γ is a circle. $\qquad\square$

Now we give the isoperimetric inequality used above.

Proposition 3.8. *The length L of a simple closed plane curve γ and the area A bounded by it satisfy*

$$L^2 \geq 4\pi A \tag{3.7}$$

with equality only for a circle.

Proof. We follow [110] and [20], [97]. Let γ be a convex and smooth, and let t, α be the coordinates in the space M of the billiard inside γ. Moreover, let $f(t,\alpha)$ be the length of the free path of the billiard ball. Consider two independent points, (t,α) and (t_1,α_1). We calculate the following non-negative integral

$$\int_{M\times M} (f(t,\alpha)\sin\alpha_1 - f(t_1,\alpha_1)\sin\alpha)^2 dt\, d\alpha\, dt_1\, d\alpha_1 = I_1 + I_2 + I_3, \tag{3.8}$$

where

$$I_1 = \int_{M\times M} f^2(t,\alpha)\sin^2\alpha_1 dt\, d\alpha\, dt_1\, d\alpha_1$$

$$= \int_0^L dt_1 \int_0^\pi \sin^2\alpha_1 d\alpha_1 \int_M f^2(t,\alpha)dt\, d\alpha = L \cdot \frac{\pi}{2} \cdot 2AL = \pi A L^2;$$

$$I_2 = -2 \int_{M\times M} f(t,\alpha)\sin\alpha f(t_1,\alpha_1)\sin\alpha_1 dt\, d\alpha\, dt_1\, d\alpha_1$$

$$= -2 \left(\int_{M \times M} f(t,\alpha) \sin \alpha \, dt \, d\alpha \right)^2 = -2(2\pi A)^2.$$

$$I_3 = I_1 = \pi A L^2.$$

Consequently, the integral (3.8) is equal to

$$2\pi A L^2 - 2(2\pi A)^2 = 2\pi A(L^2 - 4\pi A) \geq 0. \qquad \square$$

Remark 3.7. By the previous sections we know that if the domain is bounded by an ellipse (an elliptic table) then any ellipse contained in the interior of the domain with the same foci is a caustic. Moreover, in this case there exists a continuous family of caustics. But, in the case of an arbitrary convex domain such a family may fail to exist. The Kolmogorov-Arnold-Moser (KAM) theory (see [5], [65]) applied to answer the question: which plane convex billiards with smooth boundary have caustics? An application of KAM theory to plane convex billiards is due to V.Lazutkin [74], who proved that if the boundary of the domain (table) is sufficiently smooth (more precisely, the curvature of the boundary as function of the arc length should be bounded away from zero and infinity and have at least 553 continuous derivatives) then in a neighborhood of the boundary there is a one parameter family $\gamma(\eta)$, $\eta \in [0, \alpha]$, of closed curves and a set $E(\alpha) \subset [0, \alpha]$ such that $\gamma(\eta)$ is a caustic if $\eta \in E(\eta)$. Moreover, the measure of $E(\eta)$ divided by α tends to 1 as α tends to zero. The problem is connected with the problem of the existence of invariant curves of arc-preserving mappings of an annulus studied in [82].

The condition of existence of at least 553 continuous derivatives of the billiard curve, was later reduced to existence of at least 6 continuous derivatives. Lazutkin found coordinates, suggested by the string construction, in which the billiard ball map reduces to a simple form:

$$x_1 = x + y + f(x,y)y^3, \quad y_1 = y + g(x,y)y^4.$$

In particular, near the boundary of the phase cylinder $y = 0$, the map is a small perturbation of the integrable map $(x, y) \to (x + y, y)$.

3.5.2 *n-periodic trajectories on a convex table*

In this section our goal is to prove the following theorem of J. Birkhoff [39].

Theorem 3.21. *Given $n \geq 2$. For a billiard on an arbitrary plane convex domain (table) Q, with a bounded, closed, and smooth boundary ∂Q, there exists a periodic trajectory of the length n (i.e., n-periodic trajectory).*

Proof.   Let $n \geq 2$ a fixed natural number.

Case $n = 2$. In this case a periodic trajectory is given by the twice-traversed diameter AB of Q. Because, the largest chord AB in Q, is perpendicular to the tangents l_A and l_B to the boundary ∂Q at A and B, since otherwise a small shift of one of the ends A and B would increase the length of the chord AB.

Case $n \geq 3$. Let $\mathcal{P}_n$ be the set of all inscribed closed polygonal lines with number of links not exceeding n in Q. By the continuous dependence of the perimeter on the line and the compactness of the set Q there is a polygonal line $\gamma_n \in \mathcal{P}_n$ of maximal perimeter.

By the convexity of Q we have:

1) The polygonal line γ_n has exactly n links, since otherwise any link AB of it could be replaced by two links AC and CB, taking C to be a point on the arc of ∂Q between A and B, thus increasing the number of links and the perimeter of γ_n;

2) let P_{k-1}, P_k, P_{k+1} be three adjacent vertices of γ_n, then for all points P on the arc between P_{k-1} and P_{k+1} the sum of the lengths of the chords $P_{k-1}P$ and PP_{k+1} is maximal for the point $P = P_k$, because the perimeter of γ_n is maximal. Constituently, in the family of confocal ellipses with foci at P_{k-1} and P_{k+1} there is exactly one ellipse tangent to ∂Q at P_k, because of the convexity of Q.

The billiard (optical) property of ellipses implies that the segments $P_{k-1}P_k$ and P_kP_{k+1} form equal angles with the tangent l to ∂Q at P_k, moreover it coincides with the tangent to the ellipse. Thus, γ_n is a billiard trajectory, which is a n-periodic trajectory. $\square$

3.5.3   *The perimeter length function*

Let γ be a smooth strictly convex billiard curve. Assume that $x_1, \ldots, x_n \in \gamma$ are consecutive points of a n-periodic trajectory. We parameterize the curve γ by the unit circle $S^1 = \mathbb{R}/\mathbb{Z}$ so that x_i are considered as reals modulo integers.

Let us consider the space of n-gons inscribed into γ. That is the cyclic configuration space $G(S^1, n)$ that consists of n-tuples $(x_1, \ldots, x_n)$ with $x_i \in S^1$ and $x_i \neq x_{i+1}$ for $i = 1, \ldots, n - 1$.

The perimeter length of a polygon is a smooth function L on $G(S^1, n)$, and its critical points correspond to n-periodic billiard trajectories.

If one has at least two n-periodic trajectories then to distinguishes them it will be useful to use the rotation number defined as follows. Consider a configuration $(x_1, x_2, \ldots, x_n) \in G(S^1, n)$. For all i, one has $x_{i+1} = x_i + t_i$ with $t_i \in (0, 1)$; unlike x_i, the reals t_i are well defined. Since the configuration is closed, $t_1 + \cdots + t_n \in \mathbb{Z}$. This integer, which takes values from 1 to $n - 1$, is called the rotation number of the configuration and denoted by ρ.

Note that changing the orientation of a configuration replaces the rotation number ρ by $n - \rho$. Since we do not distinguish between the opposite orientations of a configuration, one assumes that ρ takes values from 1 to $\lfloor (n - 1)/2 \rfloor$.

Note that if ρ is not co-prime with n, then one may obtain an n-periodic trajectory that is a multiple of a periodic trajectory with a smaller period.

The next Birkhoff's theorem asserts that the perimeter length function has at least two extremes in each connected component of the configuration space $G(S^1, n)$. Note that its connected components are enumerated by the rotation number. Each such component is the product of S^1 and $(n - 1)$-dimensional ball.

Theorem 3.22. *Let $n \geq 2$. For every $\rho \leq \lfloor (n - 1)/2 \rfloor$, co-prime with n, there exist two geometrically distinct n-periodic billiard trajectories with the rotation number ρ.*

Proof. [110]. Similar to the case $n = 2$, one periodic trajectory is relatively easy to find. Fix a connected component M of the cyclic configuration space corresponding to the given rotation number, and consider its closure $\overline{M}$ in space $S^1 \times \cdots \times S^1$. This closure contains degenerate polygons with fewer than n sides.

Note that the perimeter length function L has a maximum in $\overline{M}$. We shall prove that this maximum is attained at an interior point, that is, not on a k-gon with $k < n$. Indeed, by the triangle inequality, the perimeter of a k-gon will increase if one increases the number of sides, see Fig. 3.58. Thus we have one n-periodic trajectory $(x_1, x_2, \ldots, x_n)$ corresponding to the maximum of L. Now to find another critical point of L in M using the minimax principle. Note that $(x_2, \ldots, x_n, x_1)$ is also a maximum point of the function L. Connect the two maxima by a curve inside $\overline{M}$ and consider the minimum of L on this curve. Take the maximum of these minima over all such curves. This is also a critical point of L, other than the maxima. A subtle point is to show that this critical point lies not on the boundary of $\overline{M}$. This follows from the fact, illustrated in Fig. 3.58, that the function L increases as one moves from the boundary. $\square$

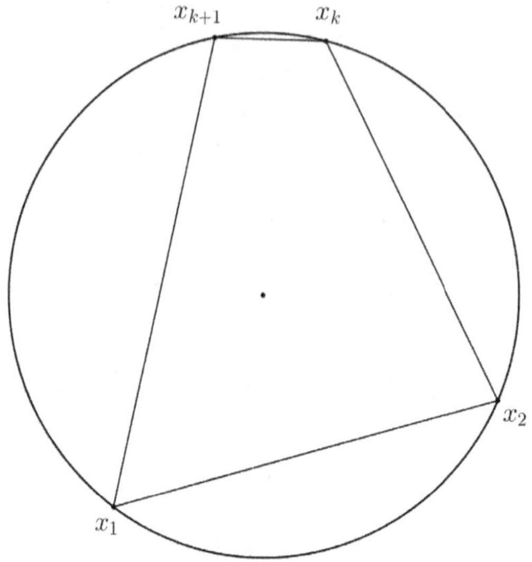

Fig. 3.58

Recall definition of polygon

Definition 3.8. A polygon is a domain $P \subset \mathbb{R}^2$ bounded by a simple closed piecewise linear curve γ. An n-gon is a polygon P whose boundary curve γ has exactly n linear pieces. These linear pieces will be called the edges of the polygon, and the singular points will be called the vertices of the polygon.

Exercise 30. Consider all possible n-gons inscribed in a circle. Find from them an n-gon with maximal perimeter.

3.6 Billiard on a polygonal table

3.6.1 *A rectangular billiard*

Following [101] we give an example of rectangular billiard. Consider a rectangular table of size $m \times n$, where m and n are integers. A ball is shot at an angle of 45 degrees from the corner of the rectangular billiard table. It always bounces off from a side at 45 degrees until it lands in a corner pocket (Fig. 3.59). We are interested to answer the following questions:

- Will it always land in a corner or will it keep bouncing forever?
- How many times will it bounce?
- Which corner pocket will it land in?

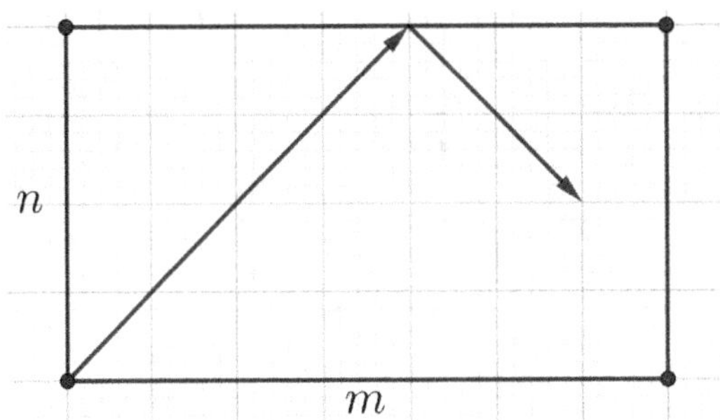

Fig. 3.59

For example, in the 7×4 table shown the ball will bounce 9 times and then fall into the top left-hand corner pocket. We shall show that the ball always ends up in a corner pocket, derive a formula for the number of bounces, and determine the finishing pocket.

Lemma 3.3. *No edge point can be hit twice.*

Proof. Assume some edge point is hit twice. Then, there must be some P that is the first edge point to be hit a second time. Let Q be the edge point hit immediately before and let R be the edge point hit immediately after the first hit of P. When P is hit a second time, the direction must be either Q to P to R, or R to P to Q. In the first case, Q is hit a second time before P is hit a second time. In the other direction, R is hit a second time before P is. Hence, in both cases, we obtain a contradiction of the statement that P is the first edge point to be hit a second time. $\square$

It is easy to see that there are at most $2m + 2n - 4$ edge-points where the ball might bounce. By Lemma 3.3 a ball visits each point at most once. Thus bouncing does not continue indefinitely, and the ball does land in a corner pocket.

To calculate the number of bounces we define the distance between two points to be the sum of the horizontal and vertical distances that separate them. For example, the opposite corners of a 4×7 table are 11 units apart.

From one bounce, A, to the next, B, how does the distance of the ball from the starting-point change? (see Fig. 3.60)

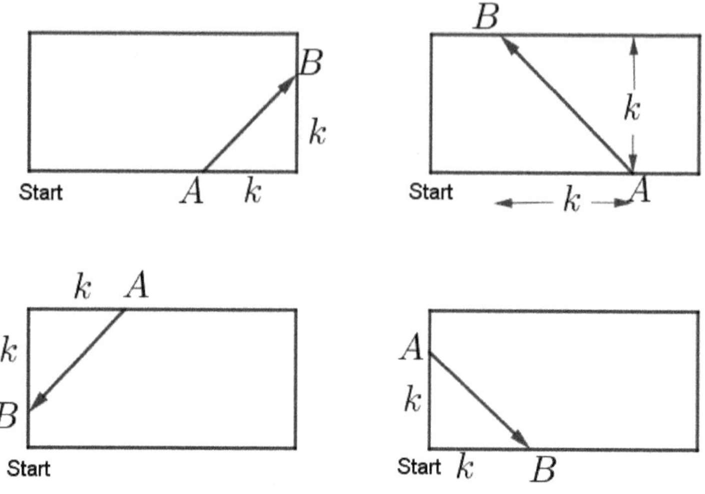

Fig. 3.60

Note that in each case the distance of B from S either equals to the distance of A from S or increases or decreases it by some $2k$.

Thus at any point where the ball bounces the distance from the start is even. This key point will be used to prove the following theorem. Consider $m \times n$ tables where the natural numbers m and n are relatively prime. Otherwise, for example, the problem of the 16×28 table is equivalent to the 4×7 one.

Theorem 3.23. *Assume m and n are relatively prime. Then the ball will bounce precisely $m + n - 2$ times before landing in a corner pocket.*

Proof. The total horizontal distance travelled by the ball from its starting corner to its finishing corner must be a multiple of m. Similarly the total vertical distance travelled must be a multiple of n. Since the angle of bounce is 45 degrees, the total horizontal and vertical distances travelled are the same. Therefore the total horizontal distance travelled is a multiple of both m and n and, hence, a multiple of mn. It is easy to see that the total horizontal distance travelled is at least mn and, therefore, that the ball must travel from one vertical edge to the other at least n times.

Thus, the ball hits a vertical edge at least $n-1$ times before reaching the corner. Similarly, the ball hits a horizontal edge at least $m-1$ times before reaching the corner and therefore hits some edge at least $m+n-2$ times. But of the $2m+2n-4$ points where the ball might bounce only half are an even distance from the start, as the reader can easily check. So by our observation that all bouncing points are an even distance from the start, a ball bounces at most $m+n-2$ times. Consequently, the ball bounces precisely $m+n-2$ times. $\qquad\square$

The last question of this subsection is: which pocket does the ball fall into? By Lemma 3.3 and the comment about even distances, it follows that the finishing-point is not the same as the starting-point and is an even distance from it. It therefore follows immediately that the finishing-points are as shown in Fig. 3.61.

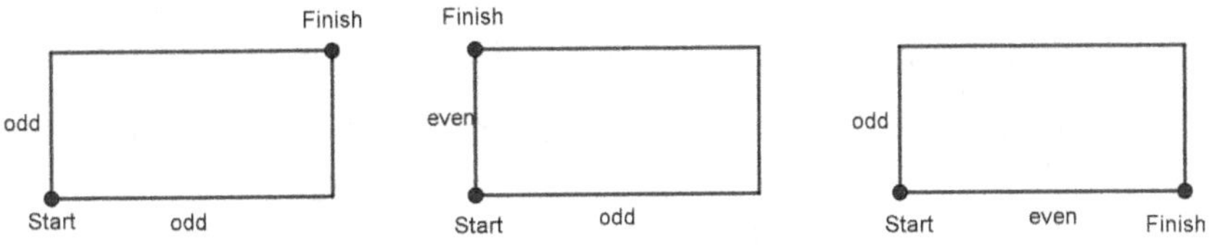

Fig. 3.61

Exercise 31. [30] The picture (see Fig. 3.62) shows some scratchwork for drawing a trajectory of slope $2/5$ (or $-2/5$) on the square torus. Starting at the top-left corner, connect the top mark on the left edge to the leftmost mark on the top edge with a line segment. Then connect the other six pairs, down to the bottom-right corner.

(a) Explain why, on the torus surface, these line segments connect up to form a continuous trajectory. Find the cutting sequence corresponding to this trajectory.

(b) Exactly where on the edges should you place the marks so that all of the segments have the same slope?

Exercise 32. In the previous exercise we put 2 marks on edge A and 5 marks on edge B and connected up the marks to create a trajectory with slope $2/5$. What if we did the same procedure with 4 marks on edge A and 10 marks on edge B?

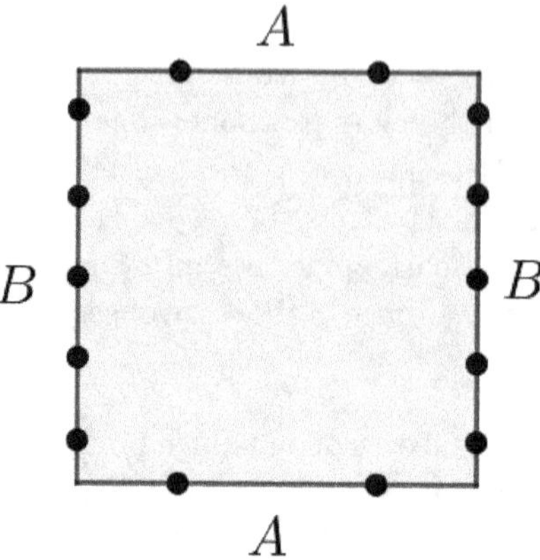

Fig. 3.62

3.6.2　*Billiard paths connecting given points*

Following [110] consider a polygonal planar domain P, and let A, B be two points inside P. We are interested to answer the following questions:

Does there exist a billiard path from A to B? This path should avoid the corners of P. This is the first illumination problem, the second being whether P can be entirely illuminated from at least one of its interior points. As shown in [112], the answer to the first question is negative. One uses very regular billiard tables to build the desired domain P.

The construction is based on the following lemma.

Lemma 3.4. *In an isosceles triangle ABC with right angle B, there is no billiard path from A coming back to A.*

Proof. Let us unfold the triangle as shown in Fig. 3.63.

The vertices labeled A, the images of the vertex A of the triangle, have both coordinates even; the vertices labelled B and C have at least one odd coordinate. If there exists a billiard trajectory in the triangle from A back to A, then its unfolding is a straight segment connecting the vertex $(0,0)$ to some vertex $(2m, 2n)$. This segment passes through point (m, n), which is either labelled B or C, or both m and n are even, and then the segment passes through point $(m/2, n/2)$, etc. Therefore, any ray starting at the vertex A and coming back to the same vertex must first pass through one of the vertices B or C of the triangle ABC, hence before hitting vertex A it must be terminated at B or C, i.e. a contradiction.　□

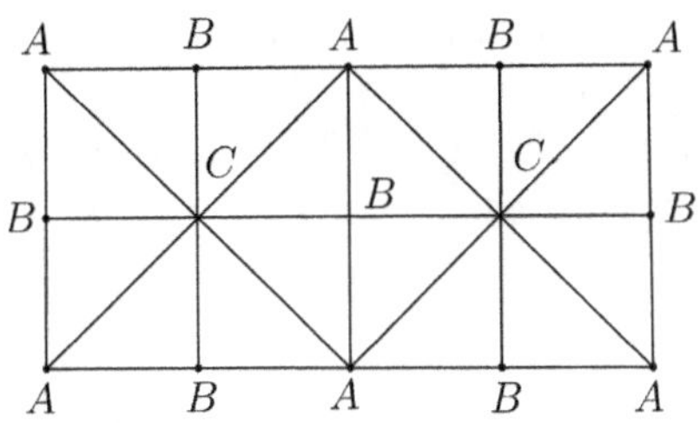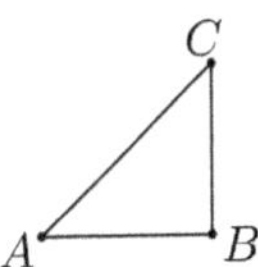

Fig. 3.63

Proposition 3.9. *For the domain P given in Fig. 3.64, there is no billiard trajectory connects points A_0 and A_1.*

Proof. The domain is constructed in such a way that all points labelled B and C are its vertices. Suppose that there exists a billiard path from A_0

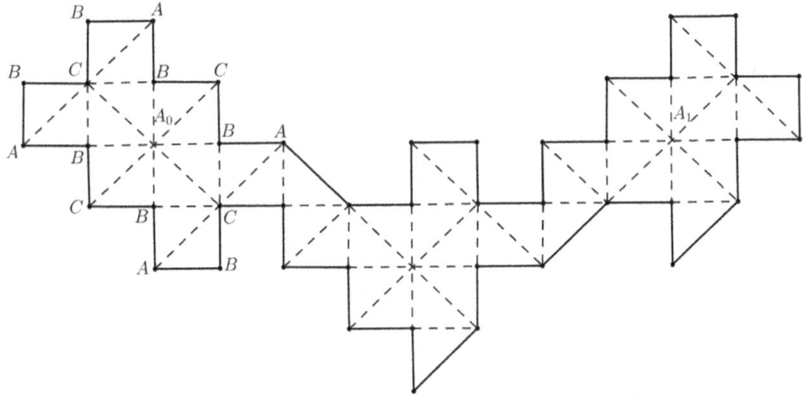

Fig. 3.64   No billiard trajectory connecting points A_0 and A_1.

to A_1. This path goes through the interior of one of the eight right isosceles triangles adjacent to point A_0. Denote this triangle by T. Then the billiard path folds down to a billiard trajectory in the triangle T that starts at A_0 and returns back to A_0. This contradicts to Lemma 3.4. $\qquad\square$

A polygon is called rational if all its angles are commensurate with π. Since the group generated by the reflections in the sides of a rational polygon is finite, we have that all links of the trajectory of a billiard ball starting at an angle φ_0 with a fixed ('horizontal') side of the polygon will give only finitely many directions with respect to this side [27]. Billiards and rational polygons were first considered in [125]. Rational polygonal billiards is a very active and fast growing area of research, see [78] for a survey of this subject.

3.6.3 *Fagnano billiard trajectories in a convex polygon*

Following [31] we give the polygonal tables admitting a shortest billiard trajectory, i.e., inscribed n-gon joining successive points on the sides. The periodic trajectories here are defined as Fagnano trajectories, which is a periodic billiard trajectory visiting each of the sides successively in order of the table exactly once before getting back to its initial point.

For convenience we denote by $\mathcal{A} = A_1 A_2 \ldots A_n$ an n-gon, with vertices numbered anticlockwise and by $I(\mathcal{A})$ denote the interior of the polygon $\mathcal{A}$, i.e. the region bounded by $\mathcal{A}$. The indices of the vertices of n-gons are tacitely considered as integers modulo n.

Definition 3.9. An n-cycle in $\mathcal{A}$ is a closed polygonal line $P_1 P_2 \ldots P_n$, where P_i lies in the open line segment $(A_i A_{i+1})$ for all i. Its length is defined by the formula

$$\lambda(P_1 P_2 \ldots P_n) = \sum_{i=1}^{n} |P_i P_{i+1}|.$$

Lemma 3.5. *Any n-cycle $\mathcal{P}$ in a convex polygon $\mathcal{A}$ is also a convex polygon.*

Proof. Note that for each $i \in \{1, 2, \ldots, n\}$, the line $\ell_{P_i P_{i+1}}$ intersects the closed polygonal region $\overline{I(\mathcal{A})}$ in the (compact) line segment $[P_i P_{i+1}]$. Let H_i be the open hyperplane bounded by the line $\ell_{P_i P_{i+1}}$ containing the vertex A_{i+1} and let H_i' the other open hyperplane. It is easy to see that, the vertices A_i and A_{i+2} are in H_i'. Suppose there exists a vertex A_j in H_i, with $j \notin \{i, i+1, i+2\}$. Since A is convex, then the vertex A_j would be a point in the triangle $\triangle A_{i+1} P_i P_{i+1}$ and thus A_j would be an interior point of the triangle $\triangle A_i A_{i+1} A_{i+2}$ which is impossible. Thus all vertices A_j, with $j \neq i$, are in the hyperplane H_i', and the same holds for all vertices P_j with $j \notin \{i, i+1\}$. Thus $\mathcal{P}$ is a convex polygon. $\square$

Definition 3.10. An n-cycle in $\mathcal{A}$ is said to be pedal if there exist a point P whose projections on the sides of $\mathcal{A}$ are exactly the points $P_1, P_2, \ldots, P_n$.

Definition 3.11. The pedal 3-cycle corresponding to the orthocenter of an acute triangle is called the orthic triangle.

Let $m(\angle ABC)$ denote the measure of an angle $\angle ABC$.

Definition 3.12. A Fagnano trajectory is an n-cycle $P_1 P_2 \ldots P_n$ in A satisfying the optic reflection law at each P_i, that is $m(\angle A_i P_i P_{i-1}) = m(\angle A_{i+1} P_i P_{i+1})$.

A polygon $\mathcal{A}$ is called cyclic if all its vertices are on a circle called the circumcircle of $\mathcal{A}$. The center of this circle is called the circumcenter of $\mathcal{A}$.

Note that the existence of a Fagnano trajectory in an n-gon $\mathcal{A}$, with $n \geq 5$, does not imply that $\mathcal{A}$ should be cyclic. Let us give an example:

Example 3.4. Consider a regular $2k$-gon $\mathcal{A} = A_1 A_2 \ldots A_{2k}$, with $k \geq 3$, and its Fagnano trajectory $F = M_1 M_2 \ldots M_{2k}$, where M_i is the midpoint of the side $(A_i A_{i+1})$. Consider also the $2k$-gon denoted by $\mathcal{A}' = A_1' A_2' \ldots A_{2k}'$, where $A_i' = A_i$ for all $i \notin \{k, k+1, k+2\}$ and the vertex A_{k+1} is translated into A_{k+1}' by a distance $\epsilon > 0$ in the direction of the vector $\overrightarrow{A_1 A_{k+1}}$ such that the line segment $A_i' A_{j+1}'$ is parallel to $A_j A_{j+1}$ for all j see Fig. 3.65. Then the $2k$-gon $\mathcal{A}'$ is not cyclic, but, at least for small positive values

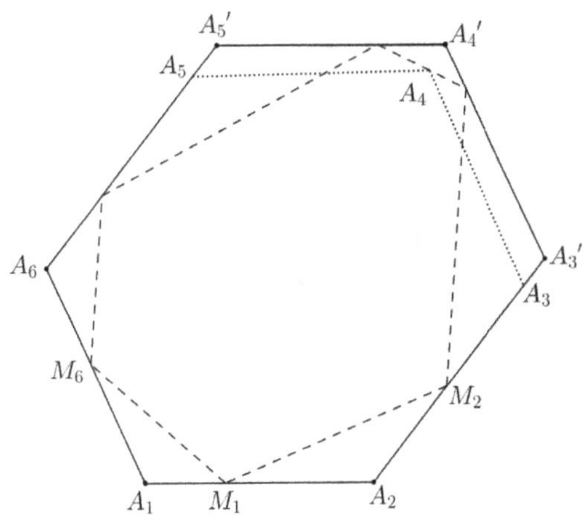

Fig. 3.65

of ϵ, it admits a Fagnano trajectory passing through the points M_i for all $i \neq k, k+1$.

Theorem 3.24. *Any n-cycle in $\mathcal{A}$ of minimal length is a Fagnano trajectory.*

Proof. Suppose that two angles $\angle P_{i-1}P_iA_i$ and $\angle P_{i+1}P_iA_{i+1}$ are not congruent for some $i \in \{1, 2, \ldots, n\}$, say

$$m(\angle P_{i-1}P_iA_i) < m(\angle P_{i+1}P_iA_{i+1}).$$

Now reflect the point P_{i+1} with respect to $\ell_{A_iA_{i+1}}$ into the point P'_{i+1} see Fig. 3.66. Then the length of the broken line $P_{i-1}P_iP'_{i+1}$ can be reduced

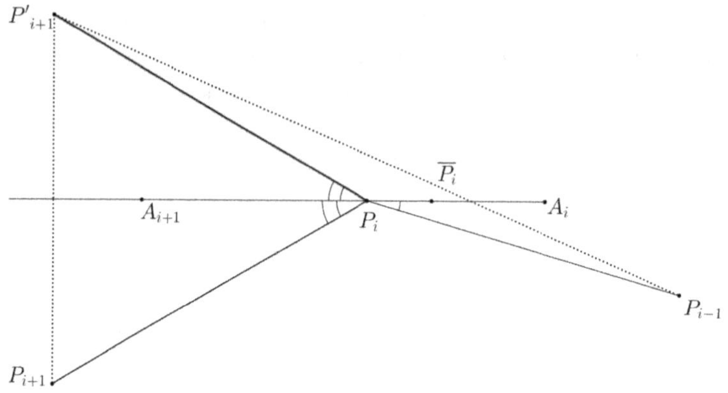

Fig. 3.66

by sliding P_i in the direction of A_i until a point $\overline{P_i}$. Thus we can obtain a new n-cycle

$$P_1P_2\ldots P_{i-1}\overline{P_i}P_{i+1}\ldots P_n$$

which is strictly shorter. $\qquad\square$

The following theorem gives a criterion to be a Fagnano trajectory

Theorem 3.25. *Let A be a cyclic n-gon with circumcenter O and let $P_1P_2\ldots P_n$ be an n-cycle in $\mathcal{A}$. Denote by θ_i the measure of the oriented angle formed by the oriented lines ℓ_{OA_i} and $\ell_{P_{i-1}P_i}$ for all i, see Fig. 3.67. Then the n-cycle $P_1P_2\ldots P_n$ is a Fagnano trajectory in $\mathcal{A}$ if and only if $\theta_i + \theta_{i+1} = \pi$ for all i.*

Proof. Follows from the following equalities

$$\theta_{i+1} = m(\angle OA_{i+1}P_i) + m(\angle P_{i+1}P_iA_{i+1}),$$
$$\pi - \theta_i = m(\angle OA_iP_i) + m(\angle P_{i-1}P_iA_i). \qquad\square$$

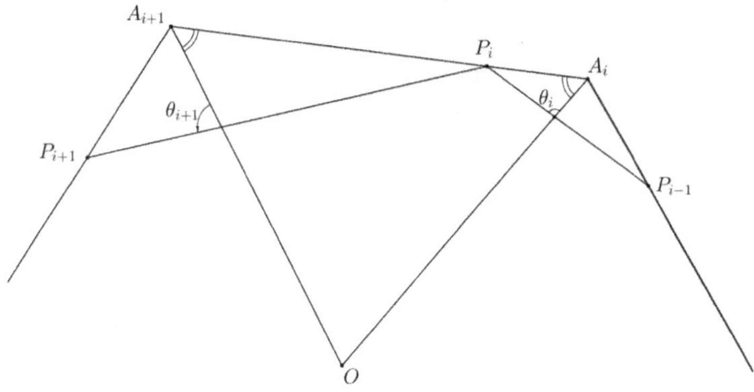

Fig. 3.67

Definition 3.13. A Fagnano trajectory in a cyclic n-gon $\mathcal{A}$ is called strong if the link $P_{i-1}P_i$ is perpendicular on the ray OA_i, for all i.

Remark 3.8. By Theorem 3.25 we get that if $\theta_{i_0} = \frac{\pi}{2}$, for some i_0 in a Fagnano trajectory in a cyclic n-gon, then $\theta_i = \frac{\pi}{2}$, for some i. Thus the Fagnano trajectory is strong. If n is odd, all Fagnano trajectories in a cyclic n-gon are strong.

Proposition 3.10. *Let $\mathcal{A} = A_1 A_2 \ldots A_n$ be a polygon. Then there exists a unique closed disk of minimal radius containing $\mathcal{A}$. Moreover if A is convex then the center O of the minimal disk is a point in $\overline{I(\mathcal{A})}$.*

Proof. Let

$$R(\mathcal{A}) = \inf\{r > 0 : \text{there is } P \text{ such that } \mathcal{A} \in \overline{B}(P,r)\}.$$

$$R(\mathcal{A}) \geq \frac{1}{2}\mathrm{diam}(\mathcal{A}) > 0.$$

Consider a sequence r_n converging to $R(\mathcal{A})$ with corresponding disks of centers P_n. By restricting ourselves to a convergent subsequence P_{n_m} we obtain a disk $\overline{B}(O, R(\mathcal{A}))$ containing $\mathcal{A}$, where $O = \lim_{m\to\infty} P_{n_m}$ which proves the existence of a minimal disk.

Suppose now $\mathcal{A} \subset \overline{B}(P, R(\mathcal{A}))$ and $\mathcal{A} \subset \overline{B}(Q, R(\mathcal{A}))$ with distance $|PQ| = 2d > 0$. Then A is contained in a ball of radius $\sqrt{R(\mathcal{A})^2 - d^2} < R(\mathcal{A})$ which is absurd, see Fig. 3.68. $\qquad\square$

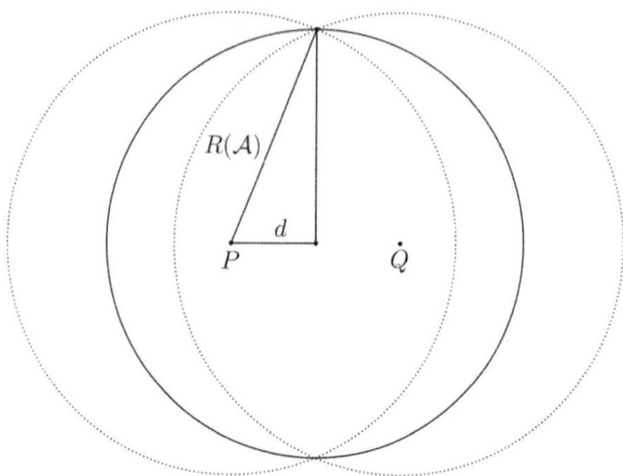

Fig. 3.68   The uniqueness of the minimal disk.

The following theorem about of cycles of extremal perimeter.

Theorem 3.26. *The length of any n-cycle $P_1 P_2 \ldots P_n$ in $\mathcal{A}$ satisfies the following inequality*

$$\lambda(P_1 P_2 \ldots P_n) \geq \frac{2\mathrm{Area}(\mathcal{A})}{R(\mathcal{A})}.$$

Moreover, if the polygon $\mathcal{A}$ admits an n-cycle $P_1 P_2 \ldots P_n$ such that

$$\lambda(P_1 P_2 \ldots P_n) = \frac{2\mathrm{Area}(\mathcal{A})}{R(\mathcal{A})},$$

then $\mathcal{A}$ is cyclic with circumcenter $O \in I(\mathcal{A})$ and the n-cycle is a strong Fagnano trajectory in $\mathcal{A}$.

Proof.   From complexity of $\mathcal{A}$ it follows that $O \in \overline{I(\mathcal{A})}$. Consider the partition of $\overline{I(\mathcal{A})}$ into the n convex quadrilaterals $OP_i A_{i+1} P_{i+1}$, for $i = 1, 2, \ldots, n$ of diagonals OA_{i+1} and $P_i P_{i+1}$ meeting under an angle $\varphi_{i+1} \in (0, \frac{\pi}{2}]$. Consequently,

$$\mathrm{Area}(\mathcal{A}) = \mathrm{Area}(\overline{I(\mathcal{A})}) = \sum_{i=1}^{n} \mathrm{Area}(OP_i A_{i+1} P_{i+1})$$

$$= \frac{1}{2} \sum_{i=1}^{n} |OA_{i+1}||P_i P_{i+1}| \sin \varphi_{i+1}$$

$$\leq \frac{1}{2} \sum_{i=1}^{n} |OA_{i+1}||P_i P_{i+1}|$$

$$\leq \frac{1}{2} R(\mathcal{A}) \lambda(P_1 P_2 \ldots P_n).$$

In case the above equality holds then we have that $P_1 P_2 \ldots P_n$ is an n-cycle in $\mathcal{A}$ of minimal length so it is a Fagnano trajectory. Since $|OA_{i+1}| = R(\mathcal{A})$ for all i, $\mathcal{A}$ is cyclic with circumcenter O. Moreover, by $\sin(\varphi_{i+1}) = 1$, for all i we get that $P_1 P_2 \ldots P_n$ is a strong Fagnano trajectory.

If $O \notin I(\mathcal{A})$ then O will lie in the open segment $(A_i A_{i+1})$ for some unique i. Using perpendicularity argument we see that the links $P_{i-1} P_i$ and $P_i P_{i+1}$ will lie in a line ℓ, cutting the convex polygon in at least three points P_{i-1}, P_i, P_{i+1}, thus ℓ must be the supporting line for some edge. Since ℓ is perpendicular on $A_i A_{i+1}$ at P_i and $P_i \in (A_i A_{i+1})$ this is absurd so the circumcenter $O \in I(\mathcal{A})$. $\qquad\square$

If $\mathcal{A}$ is a cyclic n-gon, the convex n-gon $\mathcal{B}$ whose sides are tangent to the circumcircle of $\mathcal{A}$ at the vertices of $\mathcal{A}$, successively, is called the dual of the polygon $\mathcal{A}$ (by polarity with respect to the circumcircle of $\mathcal{A}$).

The following theorem is about Fagnano trajectories bounding a maximum area.

In the proof of the next theorem we will use Lhuiliers inequality. Let us give this inequality here (see [3], pages 65-66). Let $\mathcal{A}$ be an arbitrary convex n-gon in the plane. Given a unit circle S^1 (i.e., a circle with radius 1) there exists a unique n-gon A circumscribed about S^1 such that the sides of A are parallel to the sides of $\mathcal{A}$. Denote by S and P the area and perimeter of $\mathcal{A}$. The area of A will be denoted by s.

Lhuilier's Inequality: For every convex polygon $\mathcal{A}$ we have $P^2 \geq 4Ss$, where equality holds if and only if $\mathcal{A}$ is circumscribed about a circle.

Theorem 3.27. *If $\mathcal{B}$ is the polygon dual to the cyclic n-gon $\mathcal{A}$ then for any strong Fagnano trajectory we have*

$$\mathrm{Area}(F_1 F_2 \ldots F_n) \leq \frac{\mathrm{Area}^2(\mathcal{A})}{\mathrm{Area}(\mathcal{B})}$$

where the equality holds if and only if $F_1 F_2 \ldots F_n$ is pedal.

Proof. Denote by $B_i B_{i+1}$ the side of $\mathcal{B}$ dual to the vertex A_{i+1} of $\mathcal{A}$, see Fig. 3.69. Since $F_1 F_2 \ldots F_n$ is a strong Fagnano trajectory, $F_i F_{i+1}$ and $B_i B_{i+1}$ are parallel for all i.

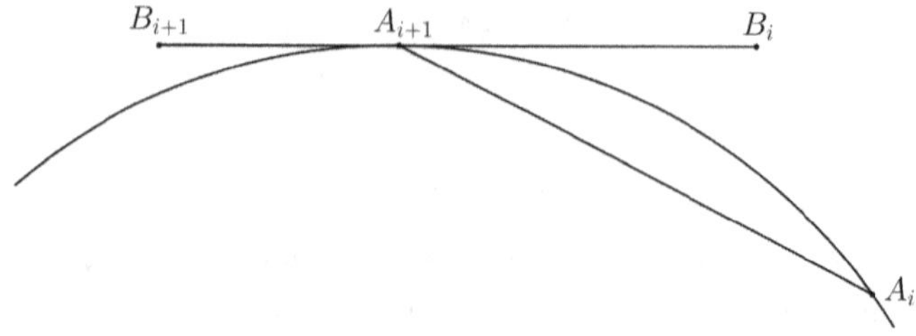

Fig. 3.69   The polygon $\mathcal{B}$ dual of $\mathcal{A}$.

By Lhuilier's inequality we have

$$4\mathrm{Area}(F_1 F_2 \ldots F_n)\mathrm{Area}(\mathcal{B}) \leq \lambda^2(F_1 F_2 \ldots F_n)R(\mathcal{A})^2 = 4\mathrm{Area}^2(\mathcal{A}),$$

where the equality holds if and only if $F_1 F_2 \ldots F_n$ admits an inscribed circle. The proof now follows from the property that a Fagnano trajectory is pedal if and only if it is circumscribed about a circle. $\qquad\square$

Remark 3.9. We note that the orthic triangle of an acute triangle is a pedal strong Fagnano trajectory. It is also easily to see that if the four projections F_1, F_2, F_3 and F_4 on the sides of a convex cyclic quadrilateral $A_1 A_2 A_3 A_4$ from the intersection point of its diagonals fall inside the respective sides, then the quadrilateral $F_1 F_2 F_3 F_4$ is a Fagnano trajectory, pedal by its construction. Moreover this quadrilateral is a strong Fagnano trajectory if and only if the cyclic quadrilateral $A_1 A_2 A_3 A_4$ considered is a Brahmagupta trapezium, which is a cyclic quadrilateral with perpendicular diagonals.[8]

Following are two constructions of Fagnano trajectories in some cyclic polygons.

Example 3.5. Let $\mathcal{A} = A_1 A_2 \ldots A_n$ be a convex cyclic n-gon containing its circumcenter. For each i, denote by B_i the midpoint of the circular arc $A_i A_{i+1}$ on the circumcircle of $\mathcal{A}$ not containing any other vertex of $\mathcal{A}$. Let $\tilde{\mathcal{A}}$ be the cyclic $2n$-gon $A_1 B_1 A_2 B_2 \ldots A_n B_n$, see Fig. 3.70. Then the polygon $\tilde{\mathcal{A}}$ admits a family of strong Fagnano trajectories. Construct a trajectory $\mathcal{F}_\epsilon$ in this family in the following way: denote

$$s = \min\{2R\cos^2 \alpha_1, 2R\cos^2 \alpha_2, \ldots, 2R\cos^2 \alpha_n\}$$

---

[8]http://mathworld.wolfram.com/BrahmaguptasTrapezium.html

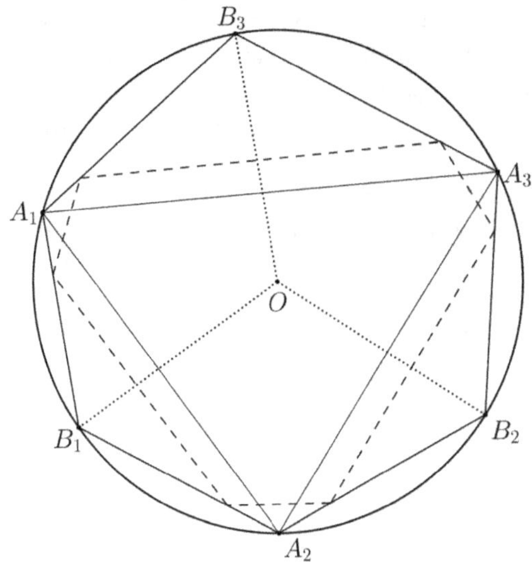

Fig. 3.70

here R is the circumradius of $\mathcal{A}$ and $2\alpha_i$ is the measure of the angle $\angle A_i B_i A_{i+1}$ for all i. For each $0 < \epsilon < s$ consider the $2n$-cycle $\mathcal{F}_\epsilon = F_1 F_1' F_2 F_2' \ldots F_n F_n'$ in $\tilde{\mathcal{A}}$ such that F_i and F_i' lie on the segments $A_i B_i$ and $B_i A_{i+1}$ respectively, and the link $F_i' F_{i+1}$ is perpendicular to OA_{i+1} for all i and its distance to the vertex A_{i+1} is ϵ, for all i. Then $\mathcal{F}_\epsilon$ is a strong Fagnano trajectory for all $0 < \epsilon < s$ (see [32]). In this family one Fagnano trajectory is pedal if and only if $\mathcal{A}$ is a regular n-gon.

An incircle of a polygon is a circle tangent to all the sides of the polygon. The center of this circle is called the incenter of the polygon.

Definition 3.14. A polygon is a Poncelet (or bicentric) polygon if it admits both a circumcircle and an incircle.

For example any triangle is bicentric. Denote $\mathrm{Ponc}_0(n)$ to be the family of all the convex Poncelet n-gons in the plane.

Definition 3.15. For $\mathcal{B} = B_1 B_2 \ldots B_n \in \mathrm{Ponc}_0(n)$ denote the contact point of the incircle of $\mathcal{B}$ on the side $B_{i-1} B_i$ by A_i, $i = 1, 2, \ldots, n$. The polygon $\mathcal{A} = A_1 A_2 \ldots A_n$ is called the contact polygon of $\mathcal{B}$.

Note that the n-gons $\mathcal{A}$ and $\mathcal{B}$ are dual by polarity with respect to the incircle of $\mathcal{B}$.

Denote by ContPonc(n) the family of all the contact polygons of convex Poncelet n-gons in the plane. It is a well-known fact that ContPonc(3) is exactly the family of acute triangles.

The following lemma is about isogonal lines in an angle.

Lemma 3.6. *Let* $\angle B_1AB_2$ *be a proper angle. Denote by* P_1 *and* P_2 *(resp. Q_1 and Q_2) the projection of two points P and Q on the lines ℓ_{AB_1}, ℓ_{AB_2} supporting the sides of the angle $\angle B_1AB_2$. Then the lines ℓ_{AP} and ℓ_{AQ} are isogonal lines[9] if and only if the quadrilateral $P_1Q_1P_2Q_2$ is cyclic. Moreover the circumcenter of $P_1Q_1P_2Q_2$ is the midpoint of the line segment $[PQ]$ (see Fig. 3.71).*

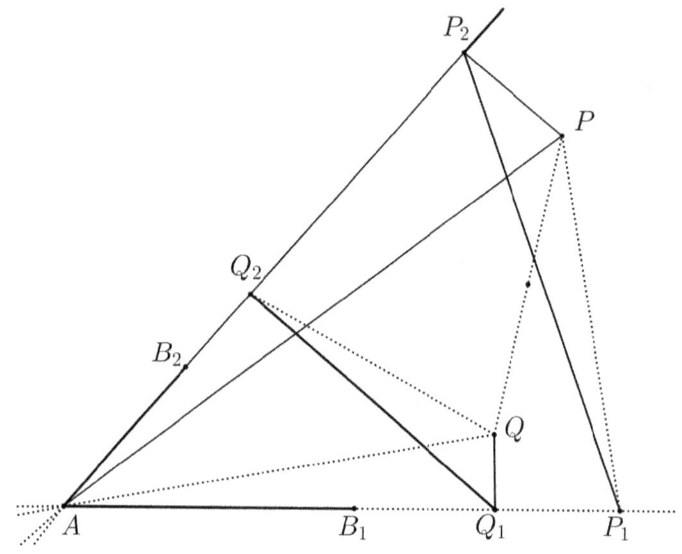

Fig. 3.71

Proof.   Denote (see Fig. 3.71)

$$\alpha_1 = m(\angle PAP_1), \quad \alpha_2 = m(\angle QAQ_2).$$

Since the quadrilaterals PP_1AP_2 and QQ_1AQ_2 are cyclic, we have

$$m(\angle P_1P_2Q_2) = \frac{\pi}{2} - m(\angle PP_2P_1) = \frac{\pi}{2} - \alpha_1,$$

---

[9]These lines are symmetrical with respect to the bisector of the angle.

$$m(\angle Q_2 Q_1 A) = \frac{\pi}{2} - m(\angle Q Q_1 Q_2) = \frac{\pi}{2} - \alpha_2.$$

Consequently, $\alpha_1 = \alpha_2$ if and only if the quadrilateral $P_1 P_2 Q_1 Q_2$ is cyclic. The midpoint of the line segment $[PQ]$, which is the intersection point of the midperpendiculars of $[P_1 Q_1]$ and $[P_2 Q_2]$ is obviously the circumcenter of the quadrilateral considered. $\qquad\square$

Theorem 3.28. *A convex cyclic polygon $\mathcal{A}$ admits a pedal strong Fagnano trajectory if and only if $\mathcal{A}$ is an element of ContPonc(n). Moreover such a trajectory in $\mathcal{A}$ is unique and it is homothetic to the Poncelet polygon $\mathcal{B}$ dual to $\mathcal{A}$.*

Proof. Let $\mathcal{A}$ be the contact of the Poncelet polygon $\mathcal{B}$. Denote by O the center of the circumcircle $\mathcal{C}$ of $\mathcal{A}$ and by M_i the midpoint of the side $A_i A_{i+1}$, see Fig. 3.72. Since the polar of the point B_i is the line $A_i A_{i+1}$, the

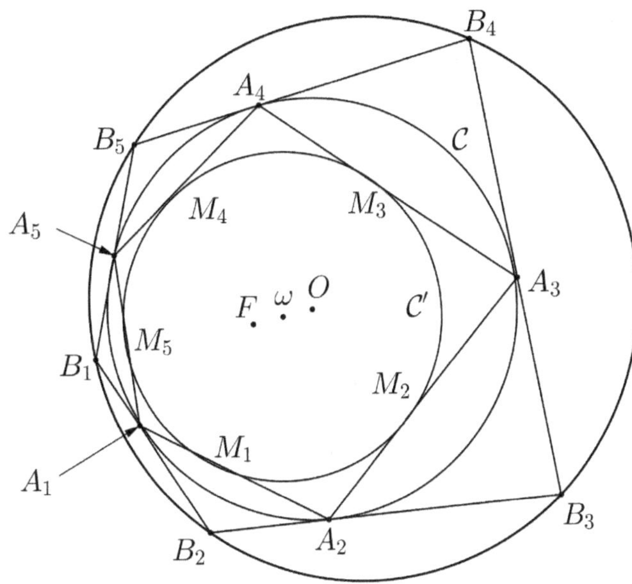

Fig. 3.72

points B_i and M_i are inverse with respect to $\mathcal{C}$. Consequently by inversion, the n-gon $M_1 M_2 \ldots M_n$ is inscribed in a circle $\mathcal{C}'$ of center ω. Moreover since the circumcircle of $\mathcal{B}$ contains the circle $\mathcal{C}$ in its interior, by inversion again, the circle $\mathcal{C}'$ is contained in the interior of $\mathcal{C}$. Hence for all i, the intersection points of the circle $\mathcal{C}'$ and the line $\ell_{A_i A_{i+1}}$ are in the open line segment $(A_i A_{i+1})$. Let F be the point such that ω is the midpoint of the line segment $[OF]$ and denote by F_i its projection on the side $A_i A_{i+1}$. We will show that $F_1 F_2 \ldots F_n$ is a pedal strong Fagnano trajectory.

Note first that for all i the line $\ell_{A_i A_{i+1}}$ and the circle $\mathcal{C}'$ intersect in exactly M_i and F_i, and thus the point F_i is in the open line segment $(A_i A_{i+1})$. Thus the n-gon $F_1 F_2 \ldots F_n$ is an n-cycle in $\mathcal{A}$, pedal by construction. Consider in the triangle $\Delta A_{i-1} A_i A_{i+1}$ the following three lines: the line $\ell_{A_i F}$, the perpendicular to $M_{i-1} M_i$ through ω, and the perpendicular to $A_{i-1} A_{i+1}$ through O (see Fig. 3.73).

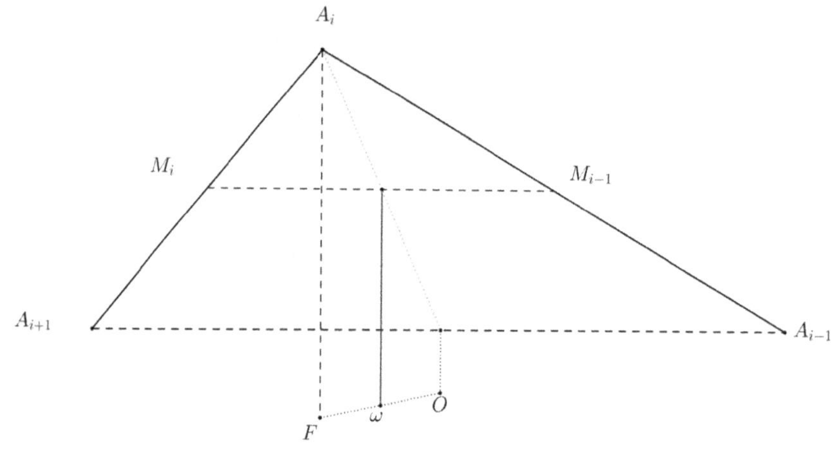

Fig. 3.73

These three lines cut both the median from the vertex A_i of the triangle $\Delta A_{i-1} A_i A_{i+1}$ and the line segment $[OF]$ in two equal parts, thus by Thales theorem, the three lines are parallel, implying that $A_i F$ is the altitude of the triangle from the vertex A_i. Consequently, $\ell_{A_i F}$ and $\ell_{A_i O}$ are isogonal lines in the proper angle $\angle A_{i-1} A_i A_{i+1}$, see Fig. 3.74. By Lemma 3.6 $M_{i-1} F_{i-1} M_i F_i$ is cyclic, therefore

$$m(\angle A_i A_{i+1} A_{i-1}) = m(\angle A_i M_i M_{i-1}) = m(\angle A_i F_{i-1} F_i).$$

It follows that $A_i O \perp F_i F_{i-1}$. Consequently $F_1 F_2 \ldots F_n$ is a strong Fagnano trajectory. In particular, since it is also pedal, the point F is its incenter. The Poncelet polygon $\mathcal{B}$ and the n-gon $F_1 F_2 \ldots F_n$ both admit an incircle,

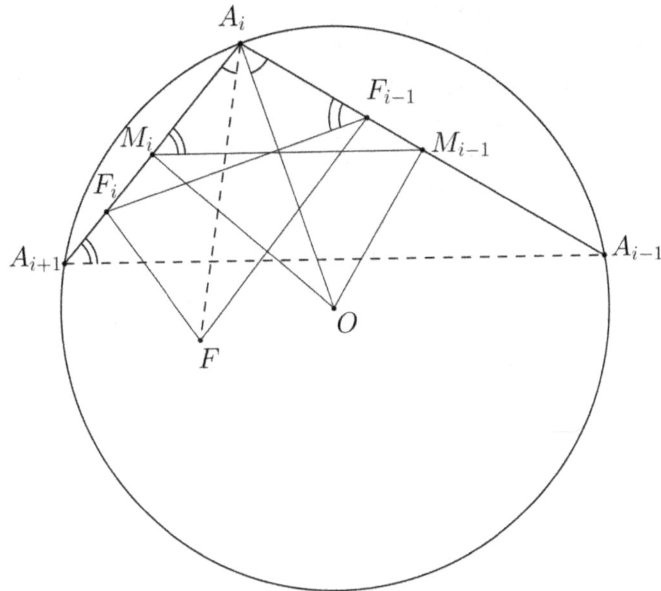

Fig. 3.74

both have their vertices numbered in increasing order anticlockwise and
have parallel corresponding sides, thus they are homothetic.

Now we shall prove the necessity of the condition in the Theorem 3.28,
let $\mathcal{F} = F_1 F_2 \ldots F_n$ be pedal strong Fagnano trajectory in the polygon $\mathcal{A}$.
By Lemma 3.5, $\mathcal{F}$ is a convex polygon and moreover all angles $\angle F_{i-1} F_i A_i$
and $\angle F_{i+1} F_i A_{i+1}$ are acute. Consequently, since the ray $O A_i$ is perpendicular to the link $F_{i-1} F_i$, for all i, it follows that the n-gon $\mathcal{A}$ contains its
circumcenter O in its interior. Let F be the unique point whose projections
on the sides are the points F_i. Thus F is the incenter of $\mathcal{F}$. Therefore
$F \in I(\mathcal{F}) \subset I(\mathcal{A})$.

Let $\alpha_i = m(A_i O A_{i+1})$ then

$$\frac{\alpha_i}{2} = m(\angle A_i A_{i-1} A_{i+1}) = m(A_i M_{i-1} M_i)$$

where M_i is the midpoint of $A_i A_{i+1}$ for all i. Since for all i, $O A_i$ is perpendicular to $F_{i-1} F_i$, we have

$$m(\angle A_i F_i F_{i-1}) = m(\angle A_i O M_i) = \frac{\pi}{2}.$$

Hence the quadrilateral $M_i F_i M_{i-1} F_{i-1}$ is cyclic. Now by Lemma 3.6 the
lines $\ell_{O A_i}$ and $\ell_{F A_i}$ are isogonal with respect to the angle $\angle A_{i-1} A_i A_{i+1}$

and the circumcenter of $M_i F_i M_{i-1} F_{i-1}$ is the midpoint of the line segment $[OF]$, denoted by ω. Thus the circumcircles of the quadrilaterals $M_i F_i M_{i-1} F_{i-1}$ are the same, for all i, say $\mathcal{C}'$. This implies that the convex n-gon $F_1 F_2 \ldots F_n$ is inscribed in a circle $\mathcal{C}'$ of center ω. Consequently the n-gon $\mathcal{F} = F_1 F_2 \ldots F_n$ is a convex Poncelet polygon. The polygon $\mathcal{F}$ and the n-gon $\mathcal{B}$ dual of $\mathcal{A}$ with respect to its circumcircle are having their sides tangent to a circle. Each are having their vertices numbered in the increasing order anticlockwise, because $O \in I(\mathcal{A})$, and their corresponding sides are parallel. Thus $\mathcal{F}$ and $\mathcal{B}$ are homothetic polygons. Consequently, $\mathcal{B}$ is a convex Poncelet polygon and $\mathcal{A}$ is an element of $\mathrm{ContPonc}(n)$. Moreover since the circumcenter ω of $M_1 M_2 \ldots M_n$ and the circumcenter O of $\mathcal{A}$ determine uniquely the point F, the uniqueness of the pedal strong Fagnano trajectory follows. $\qquad\qquad\qquad\qquad\qquad\qquad\qquad\qquad\qquad\qquad\quad \square$

Definition 3.16. The Fagnano point of a polygon $\mathcal{A} \in \mathrm{ContPonc}(n)$ is the unique point F whose pedal curve is the Fagnano trajectory $\mathcal{F}$ in Theorem 3.28.

It follows from the proof of Theorem 3.28 that:

Corollary 3.1. *The Fagnano point F and the circumcenter O are symmetrical with respect to the center ω.*

The main result of this subsection is the following

Corollary 3.2. *An n-gon is pedal strong Fagnano trajectory in some convex polygonal billiard table if and only if it is a convex Poncelet n-gon.*

Corollary 3.3. *For all $n \geq 3$, there exists infinitely many n-gonal billiard tables admitting pedal strong Fagnano trajectories.*

Corollary 3.4. *If the convex polygon $\mathcal{A}$ admits a pedal n-cycle $F_1 F_2 \ldots F_n$ such that*

$$\lambda(F_1 F_2 \ldots F_n) = \frac{2\,\mathrm{Area}(\mathcal{A})}{R(\mathcal{A})},$$

then $\mathcal{A}$ is in $\mathrm{ContPonc}(n)$ and $F_1 F_2 \ldots F_n$ is the unique pedal strong Fagnano trajectory in $\mathcal{A}$. Moreover

$$[\mathrm{Area}(\mathcal{A})]^2 = \mathrm{Area}(\mathcal{B})\,\mathrm{Area}(\mathcal{F}),$$

where $\mathcal{B}$ is the dual of $\mathcal{A}$ by polarity with respect to its circumcircle.

Theorem 3.29. *Take a polygon $\mathcal{A} = A_1 A_2 \ldots A_n$ in $\mathrm{ContPonc}(n)$.*

a) If n is odd, then $\mathcal{A}$ admits a unique strong Fagnano trajectory.

b) If n is even, then $\mathcal{A}$ admits an infinite family of parallel strong Fagnano trajectories. The unique trajectory of maximal area is the pedal trajectory $F_1 F_2 \ldots F_n$.

Proof.

a) Let $F_1 F_2 \ldots F_n$ be the unique pedal strong Fagnano trajectory in $\mathcal{A}$. Assume there exists another strong Fagnano trajectory $F_1' F_2' \ldots F_n'$. Then the oriented distance between parallel links $F_i F_{i+1}$ and $F_i' F_{i+1}'$ equals to $(-1)^i d$ for some constant d. Since n is odd and $F_1' F_2' \ldots F_n'$ is an n-cycle then d is necessarily null.

b) This family can be obtained by sliding for all i the link $F_i F_{i+1}$ along the radius OA_{i+1} an oriented distance $(-1)^i d$ for some small enough constant d. The polygonal line thus obtained closes into an n-cycle, since n is even, the cycle is a strong Fagnano trajectory. The maximal area property follows from Theorem 3.28. $\square$

3.6.4 *Periodic billiard trajectory in a polygon*

Now following [121] we shall give results answering the question: does every polygon admit a periodic billiard trajectory? This is a difficult problem. As mentioned in previous section, whether every triangle admits a periodic billiard path is still an open problem.

The best known result is that every triangle with no angle more than 100 degree will have a periodic billiard trajectory (see [98], [99]).

Definition 3.17. A labeling of an n-gon is a bijection between the set of edges of the n-gon to the set $\{1, 2, \ldots, n\}$, where adjacent edges (i.e. edges sharing a vertex) are sent to adjacent numbers in $\{1, 2, \ldots, n\}$. A polygon with a labeling is called a labeled polygon.

Definition 3.18. For an n-gon $P \subset \mathbb{R}^2$ and any billiard trajectory $\{s_i\}$ in it, let w_i be the label of the edge containing the ending point of s_i for each $i \in \mathbb{Z}$. Then $\{w_i\}$ is a sequence of labels. This is called the orbit type of the billiard trajectory.

Example 3.6. For example, the orbit type of the path in Fig. 3.75 is 1231431.

Thus the orbit type of a periodic billiard trajectory is periodic. Recall the unfolding tool for polygons: Suppose the billiard ball in a polygon hits

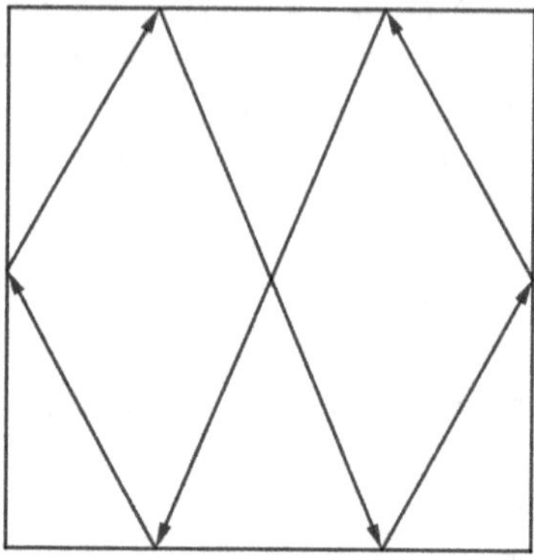

Fig. 3.75

the boundary, instead of reflecting the motion of the ball, one reflects the polygon about the edge it hits and then allow the billiard ball to go straight through. In this way, one keeps reflecting the polygon and the billiard path will be a straight line through this sequence of polygons. The obtained sequence of polygons forms an unfolding. The following is definition of unfolding for polygons.

Definition 3.19. Let $P \subset \mathbb{R}^2$ be a polygon and $\{w_i\}_{i \in \mathbb{Z}}$ be any sequence of labels of edges, let P_0 be the polygon P, and for each i, denote by P_{i+1} the polygon obtained by reflecting P_i about the edge with label w_i. Thus we inductively construct a sequence of polygons $\{P_i\}$, which is called the unfolding corresponding to the label sequence. Take their union in $\mathbb{R}^2$, i.e., $D = \cup_{i \in \mathbb{Z}} P_i$ which is the unfolding domain. If $\{w_i\}$ happens to be the orbit type for a billiard trajectory $\{s_i\}$ of P, then one also says this is the unfolding corresponding to the billiard trajectory.

Definition 3.20. Let $\{s_i\}$ be a billiard trajectory in P with corresponding unfolding $\{P_i\}$. Then each P_j has a corresponding billiard path $\{s_{i,j}\}$. Then the unfolding representation of $\{s_i\}$ is the union $L = \cup_{i \in \mathbb{Z}} s_{i,i}$, which would be a straight line contained in the unfolding domain D.

Proposition 3.11. *If in a polygon P, two billiard trajectories $\{s_i\}$ and $\{t_i\}$ have the same orbit type, then s_i is parallel to t_i for all $i \in \mathbb{Z}$.*

Proof. Let $\{P_i\}$ be the corresponding unfolding of the trajectory, and let D be the unfolding domain. Then one can rotate and translate everything so that the unfolding representation of $\{s_i\}$ is the straight line coinciding with the x-axis. Denote by $L \subset D$ the unfolding representation of $\{t_i\}$. Note that P_i is the same polygons with the same size for all $i \in \mathbb{Z}$, so let $d > 0$ be the diameter of P_i for all i. Assume L is not horizontal. Then it can be parameterized by $x = ay + b$ for some $a, b \in \mathbb{R}$. Take a point (x_0, y_0) on L with $y_0 > d$ and $x_0 = ay_0 + b$. Then as $L \in D$, $(x_0, y_0) \in P_i$ for some $i \in \mathbb{Z}$. However, as P_i has diameter d and it intersects with the positive x-axis, all points in P_i will have y coordinates less than or equal to d, contradiction. Therefore L is parallel to the x-axis. This completes the proof. $\qquad\square$

By this proposition we have that an orbit type determines the billiard trajectory up to a translation. So one can classify billiard trajectories first by their orbit types.

Proposition 3.12. *In a polygon P, a billiard trajectory is periodic if and only if its orbit type is periodic.*

Proof. The necessity is clear. To prove sufficiency, let $\{s_i\}$ be any billiard trajectory with periodic orbit type $\{w_i\}$ with minimal period p. Define k as follows

$$k = \begin{cases} p, & \text{if } p \text{ is even} \\ 2p, & \text{if } p \text{ is odd,} \end{cases}$$

therefore, k is always even. This number k is the minimal even period. Let $\{P_i\}$ be the unfolding corresponding to $\{w_i\}$, D be the unfolding domain, and L be the unfolding representation of $\{s_i\}$. Now for each P_i, let c_i be its centroid. Since k is even, P_0 and P_k should have the same orientation, i.e. we can obtain P_k from P_0 by a translation plus rotation. Let $\tau : \mathbb{R}^2 \to \mathbb{R}^2$ be the translation, and $r : \mathbb{R}^2 \to \mathbb{R}^2$ be the rotation centered on c_k such that $f = r \circ \tau$ will send P_0 to P_k. Then by periodicity, f^n would send P_0 to P_{nk}.

Now we connect the points $\ldots c_{-2k}, c_{-k}, c_0, c_k, c_{2k}, c_{3k}, \ldots$ by line segments. Assume r is not a rotation by a multiple of 2π, then $\angle c_0 c_k c_{2k}$ is not a multiple of 2π. Then the points c_0, c_k, c_{2k} will determine a circle S. Moreover, in case $\angle c_0 c_k c_{2k}$ is an odd multiple of 2π, c_0 and c_{2k} will coincide, so S will be the circle with diameter $c_0 c_k$. Since f is a rigid motion and $f^n(c_0 c_k) = c_{nk} c_{(n+1)k}$ and $f^n(\angle c_0 c_k c_{2k}) = \angle c_{nk} c_{(n+1)k} c_{(n+2)k}$, one can see that c_{nk} lies on S for all $n \in \mathbb{Z}$. Let C be the center of S, and let

$$d = \sup\{|p - C| : p \in \cup_{i=0}^{k} P_i\},$$

the existence of which follows from the fact that $\cup_{i=0}^{k} P_i$ is compact. Then by periodicity we have

$$d = \sup\{|p - C| : p \in \cup_{i\in\mathbb{Z}} P_i = D\}.$$

Denote by S' the closed ball centered at C with radius d, then $D \subset S'$ must be bounded. However, the straight line L is contained in D and cannot be bounded, which is a contradiction. Therefore, r must be a rotation by a multiple of 2π. Hence P_0 and P_k differ only by a translation. The segment s_0 ends in the edge w_0, and s_k ends in the edge $w_k = w_0$ in the same direction. Now we rotate and scale everything so that the edge w_0 of P is the line segment from $(0,0)$ to $(1,0)$. Assume s_0 ends in $(a,0)$ and s_k ends in $(b,0)$, and suppose $a \neq b$. Without loss of generality, we take $b > a$. Then by periodicity of the unfolding, s_{2k} ends in $(b + (b-a), 0)$, and s_{nk} would end in $(a + n(b-a), 0)$. For n large enough, we would have $a + n(b-a) > 1$. Then s_{nk} would be out of the polygon P_{nk}, a contradiction. Consequently, we can only have $a = b$. So s_0 and s_k would be line segments in P ending in the same spot with the same direction. So the billiard path would repeat itself. Thus the billiard path is periodic. This completes the proof. $\qquad\square$

We have the following

Corollary 3.5. *If a billiard trajectory is periodic, let $\{P_i\}$ be the corresponding unfolding, and let p be the minimal even period. Then P_p can be obtained from P_0 by a translation, and this translation is in the direction of the unfolding representation L of the billiard trajectory.*

Exercise 33. Numerate the edges of a rectangle by $1,2,3,4$. Consider a trajectory of billiard in the rectangle and write the sequence $n_1, n_2, \ldots$ ($n_i \in \{1,2,3,4\}$) of numbers of edges in order of reflections of the trajectory from the edges. Prove that the trajectory is periodic if and only if the sequence $\{n_i\}$ is periodic.

For a given polygon, there are many possibilities to label its edges, and different ways of labeling will give the same billiard trajectory different orbit types. Therefore, to study billiard trajectories by the orbit type, it would be convenient to have a space of labeled n-gons. This is given in the following

Definition 3.21. The space of labeled n-gons is

$$\mathcal{P}_n = \{n - \text{gons in } \mathbb{R}^2 \text{ with a labeling}\}.$$

The following proposition gives structure of the space:

Proposition 3.13. *The set $\mathcal{P}_n$ is an open subset of $\mathbb{R}^{2n}$.*

Proof. Let $f : \mathcal{P}_n \to \mathbb{R}^{2n}$ be a map such that for each labeled polygon P, let $v_i \in \mathbb{R}^2$ be the vertex between edges i and edges $i+1$, then $f(P) = (v_1, \ldots, v_n) \in \mathbb{R}^{2n}$. Since each polygon is determined by its vertices, and the ordering of its vertices determines its labeling, we see that this map is injective. So $\mathcal{P}_n$ can be seen as a subset of $\mathbb{R}^{2n}$. If the collection of n vertices $v = (v_1, \ldots, v_n)$ form a labeled polygon, let $r = \frac{1}{2}\min_{i,j}|v_i - v_j|$. Let B_i be the open ball in $\mathbb{R}^2$ with center v_i and radius r, and consider the open set $R = \cap B_i$. For any $p = (p_1, \ldots, p_n) \in R$, then $p_i \in B_i$ for all i, and consider the polygonal curve γ by joining $p_1 p_2, \ldots, p_n p_1$. This is a closed piecewise linear curve. If $v_i v_{i+1}$ and $v_j v_{j+1}$ are disjoint line segments, then the vertices of these two line segments are at least $2r$ units apart, so points on one of the line segments will be at least $2r$ apart from points on the other. Therefore $p_i p_{i+1}$ and $p_j p_{j+1}$ cannot intersect. So γ indeed determines a polygon P. We give the edge with vertices p_n, p_1 the label 1, and give the edge with vertices p_i, p_{i+1} the label $i+1$. Then $p \in \mathcal{P}_n$ and $R \subset \mathcal{P}_n$. Consequently, $\mathcal{P}_n$ is an open set. $\qquad\square$

Note that congruent labeled polygons will have the same configuration of billiard trajectories.

Definition 3.22. A function $f : \mathbb{R}^2 \to \mathbb{R}^2$ that is the composition of a rigid motion and a scaling is called a congruence map.

For any unlabeled polygons P and P', if $f : P \to P'$ is a congruence map, then f would induce a bijection between edges of P and edges of P'. Let $\ell : \{\text{edges of } P\} \to \{1, 2, \ldots, n\}$ be any labeling for P, then one has an induced labeling for P', defined by $f_* \ell(a) = \ell \circ f(a)$ for each edge a of P'.

Definition 3.23. Two labeled polygons $P, P' \in \mathcal{P}_n$, with labeling ℓ, ℓ' respectively, are called congruent if there exists a congruence map $f : P \to P'$ such that $f_* \ell = \ell'$.

Proposition 3.14. *The congruence is an equivalence relation.*

Proof. *Reflexivity.* Note that all $P \in \mathcal{P}_n$ is congruent to itself through the identity map.

Symmetricity. Since rigid motions are invertible with inverse again a rigid motion, and the same is true for scaling, we conclude that a congruence map will have an inverse a congruence map. Thus if P is congruent to P' through map f, then P' is congruent to P through map f^{-1}.

Transitivity. Clearly, if P is congruent to P' through map f, and P' is congruent to P'' through map g, then P is congruent to P'' through map $g \circ f$. $\square$

Proposition 3.15. *If $P, P' \in \mathcal{P}_n$ are two congruent polygons. Then P has a billiard trajectory with an orbit type $w = \{w_i\}$ if and only if P' has a billiard trajectory with the orbit type w.*

Proof. Note that if $\{s_i\}$ is a billiard trajectory for P with orbit type w, then $\{f(s_i)\}$ would be a billiard trajectory for P'. Moreover, the ending point of $f(s_i)$ lies on the edge whose preimage on P has label w_i. Then this edge will also have label w_i. Consequently $\{f(s_i)\}$ will have orbit type w as well. The other direction follows from the fact that congruence is an equivalence relation. $\square$

By this proposition it is sufficient to work only on congruence classes of labeled polygons and ask if they have billiard paths of certain orbit type. Thus the main objects of interests are the following.

Definition 3.24. The space of labeled n-gons modulo congruence is $\tilde{\mathcal{P}}_n = \mathcal{P}_n / \sim$, where $\sim$ is the congruence relation among n-gons, and endorse $\tilde{\mathcal{P}}_n$ with the quotient topology.

Proposition 3.16. *The set $\tilde{\mathcal{P}}_n$ is an open subset of $\mathbb{R}^{2n-4}$.*

Proof. For each congruence class, pick any representative with vertices $(v_1, \ldots, v_n)$. Then we translate so that $v_1 = (0,0)$, and rotate and scale so that $v_2 = (1,0)$, and we finally reflect so that v_3 is in the upper half plane. This gives a map from $\tilde{\mathcal{P}}_n$ to an open subset of $\mathbb{R}^{2n-4}$. It is trivial to check that this map is a well-defined embedding. $\square$

Definition 3.25. For a given orbit type $w = \{w_i\}$, its orbit tile $P(w) \subset \tilde{\mathcal{P}}_n$ is the set of all congruence classes of labeled n-gons which have a billiard path with orbit type w. An orbit type is stable if its orbit tile is an open set. A periodic billiard path is stable if its orbit type is stable.

Definition 3.26. Given a polygon P and a periodic orbit type $\{w_i\}$ with minimal even period p, let $\{P_i\}$ be the corresponding unfolding. For each P_i, let $bdry_i$ be the union of all its vertices and all its edges with edge-label different from w_{i-1} or w_i. The boundary of the unfolding domain D is the union of all $bdry_i$.

Proposition 3.17. *The boundary of the unfolding domain is always the union of two unique piecewise linear curve.*

Proof. For P_0, it is easy to see that $bdry_0$ has two connected components, both are piecewise linear curves, here one treats a single vertex as a piecewise linear curve. Let us denote these curves as u_0 and ℓ_0. We use mathematical induction. Assume now $\cup_{i=-n}^{n} bdry_i$ is the union of two disjoint piecewise linear curves, let the one containing u_0 be u_n and let the one containing ℓ_0 be ℓ_n. Then for P_{n+1}, the boundary $bdry_{n+1}$ also has two connected components, both piecewise linear, and they are connected to u_n, ℓ_n through the two vertices of w_n edge of T_n respectively. Therefore we can extend u_n and ℓ_n to include these two components respectively. Since no piece in $bdry_{n+1}$ can contain both vertices of w_n, this extension is unique. Similarly we also extend them uniquely to include the two pieces in $bdry_{-n-1}$. By this way we obtain u_{n+1} and ℓ_{n+1}. Thus the induction completes the proof. $\square$

Definition 3.27. For two piecewise linear curves $\gamma, \gamma' \subset \mathbb{R}^2$, we say that they are separable if there is a straight line $L \subset R^2$ such that γ and γ' are contained in the two distinct connected components of $\mathbb{R}^2 \setminus L$.

The following proposition is about existence of periodic billiard trajectory

Proposition 3.18. *If P is an arbitrary polygon, given any periodic orbit type $w = \{w_i\}$ with minimal even period p and unfolding $\{P_i\}$, then P admits a billiard trajectory with orbit type w if and only if the two piecewise linear curves in the boundary of the unfolding domain are separable.*

Proof. *Sufficiency.* Let L be the unfolding representation of the billiard trajectory $\{s_i\}$ with orbit type w. Then L is a straight line contained in the unfolding domain D. Since L only touch the edge w_{i-1} and w_i of P_i, it is disjoint from the two piecewise linear curves in the boundary of the unfolding domain, and since $L \subset D$, the two curves of D must be on different side of L, i.e. L separates the two piecewise linear curves in the boundary of the unfolding domain. So sufficiency is proved.

Necessity. Let L be the line separating the two piecewise linear curves in the boundary of the unfolding domain D. Then it is easy to see that $L \subset D$, and for each P_i, we have that L intersects P_i only with its w_{i-1} and w_i edges. Since P_i and T are congruent as labeled polygons, we let $f_i : P_i \to P$ be a congruence map. Denote $t_i = L \cap P_i$, and $s_i = f_i(t_i)$ for all i. We shall prove that $\{s_i\}$ is a periodic billiard path in P with orbit type

w_i. Indeed, since f_i are label preserving, each s_i start in edge w_{i-1} and ends in edge w_i. Since P_i and P_{i+1} differs only by reflection $r : P_{i+1} \to P_i$ about w_i edge, thus we say that $f_{i+1} = f_i \circ r$. As r is a reflection, and t_i, t_{i+1} are on the same line L and they connect at a single point on w_{i+1}, we see that t_i and $r(t_{i+1})$ hit the same point on the w_{i+1} edge of P_i, and their angles with the w_{i+1} edge are complementary. Then the same must be true for $f_i(t_i) = s_i$ and $f_i(r(t_{i+1})) = f_{i+1}(t_{i+1}) = s_{i+1}$. Consequently, $\{s_i\}$ is a periodic billiard trajectory with orbit type w. $\qquad\square$

Example 3.7. The above proposition is nicely exemplified by Fig. 3.76.

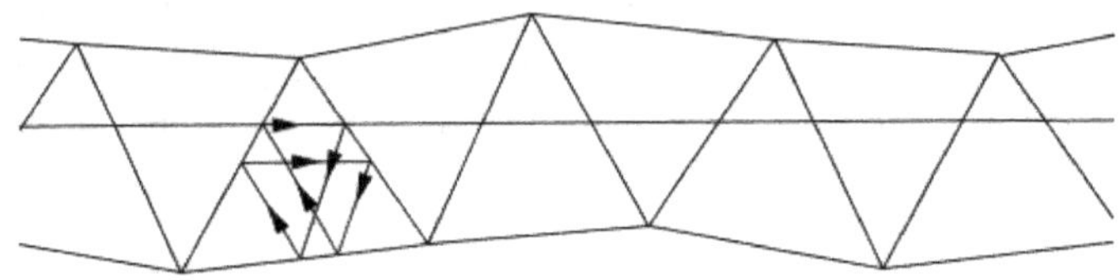

Fig. 3.76   The unfolding of the triangle according to the orbit type 123123. Exemplifying Proposition 3.18.

Now we discuss stable periodic billiard trajectories and the unstable ones. Recall that stable billiard trajectories are those that survive any small perturbation of the polygon. So it is more effective to use stable periodic orbits to cover $\tilde{\mathcal{P}}_n$. There are many polygons without stable periodic billiard trajectory, e.g. all right triangles [58]. Therefore an understanding of the unstable ones is also important.

Proposition 3.19. *Suppose an unstable periodic orbit type with nonempty orbit tile is given, assuming the relation $\sum_{i=1}^{n} \theta_i = (n-2)\pi$, where θ_i are all the inner angles. There exists a unique nontrivial linear relation over $\mathbb{R}$ on inner angles such that any n-gon whose congruence class is in that orbit tile will satisfy this linear relation. Moreover, the coefficients of this relation can be taken to be in $\mathbb{Z}$. On the other hand, for a stable periodic orbit type with nonempty orbit tile, there is no such linear relation.*

Proof.   Consider an arbitrary periodic orbit type. For any n-gon P in its orbit tile, let θ_i be the angle between edges i and $i+1$ for all $i \in \{1, 2, \ldots, n\}$.

Denote

$$[s,t] = \sum_{i \in \{s, s+1, \ldots, t-1\}} \theta_i.$$

The i-th vertex would be the vertex for the angle θ_i. Let $w = \{w_1, \ldots, w_{2p}\}$ be a single minimal even period for the given periodic orbit type, and let $\{P_i\}$ be the unfolding for the orbit type. For each $k = 1, 2, \ldots$, the polygon P_{2k+1} is obtained from P_{2k-1} by reflecting through the edge w_{2k-1} and then reflecting through the edge w_{2k}. So from P_{2k-1} to P_{2k+1} would be a translation plus a rotation of degree $2[w_{2k-1}, w_{2k}]$. Since P_1 to P_{2p+1} is just a translation as in Corollary 3.5, we have $\sum_{k=1}^{p} 2[w_{2k-1}, w_{2k}] = 0$. This is a linear relation with even integer coefficients on inner angles, and it is satisfied by all n-gons in the orbit tile of the given orbit type. If this is trivial or equivalent to the relation that the sum of inner angles is $(n-2)\pi$, then for all n-gons P, let $\{P_i\}$ be the corresponding unfolding. Then from P_0 to P_{2p} is a translation. Let $\mathbf{P}$ be any polygon whose congruence class is in the orbit tile. Let D be its unfolding domain. Then by Proposition 3.18 the boundary of the unfolding domain is the union of two separable piecewise linear curves. Let L be such a separating line, and let $r > 0$ be the minimal distance from any point of the boundary of the unfolding domain to the line L. Let Q be a polygon such that the i-th vertex of it is less than $\epsilon > 0$ away from the i-th vertex of $\mathbf{P}$ for all i, where ϵ is a small value to be determined. Note that the set of all such n-gons forms an open neighborhood of $\mathbf{P}$. We translate, rotate and scale Q to obtain Q' such that the centroid of Q' is the same as $\mathbf{P}$, and in the unfolding the centroid of Q'_{2p} is the same as $\mathbf{P}_{2p}$. Note that the amount of translation and rotation for ϵ small enough can be made arbitrarily small. Consequently, the i-th vertex of Q' can be made arbitrarily near to the i-th vertex of $\mathbf{P}$. Now once corresponding vertices of $\mathbf{P}_0, Q'_0$ can be made arbitrarily near, then the corresponding vertices of $\mathbf{P}_1, Q'_1$ can also be made arbitrarily near, and so forth. Now we set $\epsilon > 0$ so small that the corresponding vertices of $\mathbf{P}_i$, and Q'_i are less than r for all $i \in \{0, 1, \ldots, 2p-1\}$. This is possible since we only required this for finitely many i. Now by periodicity, corresponding vertices of $\mathbf{P}_i$, and Q'_i are less than r away for all $i \in \mathbb{Z}$. Then the two piecewise linear curves in the boundary of unfolding domain for Q' will never touch the line L, and L will be in this unfolding domain. Then by periodic billiard path existence lemma, Q_0 has a periodic billiard path with the given orbit type, and so does Q. Consequently, the orbit tile for the orbit type is open and stable. If the linear relation $\sum_{k=1}^{p} 2[w_{2k-1}, w_{2k}] = 0$ is nontrivial and not equivalent with the relation that the sum of inner angles is $(n-2)\pi$. Then this linear relation gives a hypersurface of $\mathbf{P}$, and as the orbit tile is a subset of this hypersurface, it cannot be open. Consequently the orbit type is unstable. Moreover, by similar arguments as above we can see that the orbit tile is an open subset of this hypersurface.

Finally, suppose there is any other linear relation over $\mathbf{R}$ on inner angles satisfied by all polygons in the orbit type. If the orbit type is stable, this means that the orbit tile is contained in a hypersurface, and thus not open, a contradiction. If the orbit type is unstable, then this means that the orbit tile is contained in two distinct hypersurfaces. Then their intersection cannot be open in each of the hypersurfaces, again a contradiction. This completes the proof. $\qquad\square$

Definition 3.28. The linear relation in the above proposition is called the canonical linear relation for the orbit type.

Definition 3.29. A finite sequence of labels $(w_1, \ldots, w_p)$, where $w_i \in \{1, \ldots, n\}$, is said to be well-balanced if for each $k \in \{1, \ldots, n\}$, we have

$$|\{w_i : i \text{ is even and } w_i = k\}| = |\{w_i : i \text{ is odd and } w_i = k\}|.$$

For example, the sequence 123123 is well-balanced, but 12341234 is not.

Proposition 3.20. *A periodic orbit type is stable if and only if a single minimal even period of it is well-balanced.*

Proof. Let $f(\theta_1, \ldots, \theta_n) = 0$ be the canonical linear relation for the orbit type $w = \{w_i\}$ with minimal even period $2p$, and let $w' = \{w_1, \ldots, w_{2p}\}$. We shall show that, assuming the relation $\sum_{i=1}^{n} \theta_i = (n-2)\pi$, w' is well-balanced if and only if $f \equiv 0$. For any $a, b, c \in \{1, 2, \ldots, n\}$, we have by definition that

$$[a, b] + [b, c] = [a, c] + S$$

where S is a multiple of the sum $\sum_{i=1}^{n} \theta_i$. This sum is a multiple of π, so $2S$ is a multiple of 2π, and thus $2S = 0$. Hence $2[a, b] + 2[b, c] = 2[a, c]$. Consequently, if w' is well balanced, we have $f \equiv 0$. On the other hand, assume $f \equiv 0$. Let $[a, b]$ be an interval in f, i.e. $[a, b] = [w_{2k-1}, w_{2k}]$ for some k. Now θ_{b-1} is contained in an even number of intervals, and so does θ_b, but $[a, b]$ does not contain θ_b any more. Therefore, there exists an interval $[b, c]$ of f. Because f has finitely many intervals, continuing this process, we can eventually find an interval ending in a, and all these intervals we find sums up to 0. Now we throw these intervals away from f as they sum up to 0. If there is no other interval left, then clearly w_0 is well balanced. Otherwise we pick any interval left, and repeat the above process. Since f has finitely many intervals, eventually there will be no interval left, and we see that w' is well-balanced. $\qquad\square$

We have the following

Corollary 3.6. *If an orbit type has an odd minimal period, then it is stable.*

We finish this section with some results related to rhombi and rectangle following [121]. We do not give proofs of these results, because they are based on 'heavy' (for this book) subjects and techniques (see [121] for proofs).

Consider the rhombi with one angle equal to $\frac{\pi}{n}$, $n \geq 2$.

Theorem 3.30. *A rhombi R_n with on angle $\frac{\pi}{n}$ has no stable periodic billiard trajectory when n is a power of 2.*

Corollary 3.7. *A square has no stable periodic billiard trajectory.*

Now consider rectangles.

Conjecture.[121] All parallelograms with one angle $\frac{\pi}{2n}$ for n a power of 2 will have no stable periodic billiard trajectory.

Consider the case $n = 1$, in which case the parallelogram must be a rectangle.

Theorem 3.31. *A rectangle R has no stable periodic billiard trajectory.*

Definition 3.30. Choose an edge of a parallelogram as the base. The modulus of the parallelogram is the ratio of its base and its height.

Note that by different choice of base edge (see Fig. 3.77), we have two moduli for each parallelogram.

Corollary 3.8. *A billiard path in a rectangle is periodic if and only if its slope is a rational multiple of the modulus, and it's trajectory is dense in the rectangle if and only if its slope is an irrational multiple of the modulus.*

Now we can move on to parallelogram with one angle $\frac{\pi}{4}$.

Theorem 3.32. *A parallelogram P_1 with one angle $\frac{\pi}{4}$ and modulus 1 have no stable periodic billiard trajectory.*

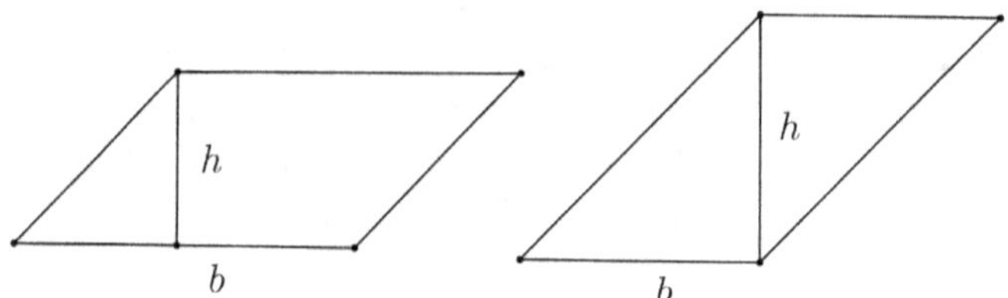

Fig. 3.77　The two moduli for a parallelogram. Here b is the base and h is the height, and their ratio is the modulus.

A criterion for existence of a periodic trajectory in a regular hexagon is given in the following theorem.

Theorem 3.33. [118] *A billiard trajectory γ in a regular 6-gon is periodic if and only if one (and then also all) of its links forms an angle*

$$\alpha = \arctan \frac{1}{\sqrt{3}(2(x/y) - 1)}$$

with the side with which it has a point in common; here, x and y are coprime integers and $y > 0$. Moreover, a periodic trajectory γ launched at this angle α from a side of the hexagon is stable if and only if it is invariant under rotations of the hexagon through $120°$ and y is divisible by 3.

Using a wide ribbon whose center line is the trajectory, wrapping around the triangle one can solve the following exercises which mean that a periodic trajectory on a polygonal billiard table is never isolated (see [30]):

Exercise 34. Prove that each even-periodic trajectory contained in a 1-parameter family of parallel periodic trajectories of the same period and length.

Exercise 35. Show that each odd-periodic trajectory belongs to a family of trajectories whose period and length is twice as great.

Exercise 36. For a billiard on the square table, solve the following exercises.

1. How many 10-periodic trajectories do exist on this table?
2. How many 14-periodic trajectories do exist? Draw each of these trajectories.
3. Show that a trajectory with slope p/q on the square billiard table has period $2(p + q)$.

4. Let p be a natural number, how many $2p$-periodic billiard trajectories (up to symmetry) are there?

5 How many periodic trajectories of length less than L^{10} do exist? How long is the trajectory of slope 2? The trajectory of slope 1/3?

For the square table, it is known that every trajectory is either periodic, or its path eventually covers the table. Does this always happen? Surprisingly, the answer is no! There are tables where some trajectory completely covers one region of the table, but never goes to another region of the table: An illustration of this phenomenon, constructed by Curtis McMullen (see [29]), is shown in Fig. 3.78. The trajectory gets "trapped" in the rectan-

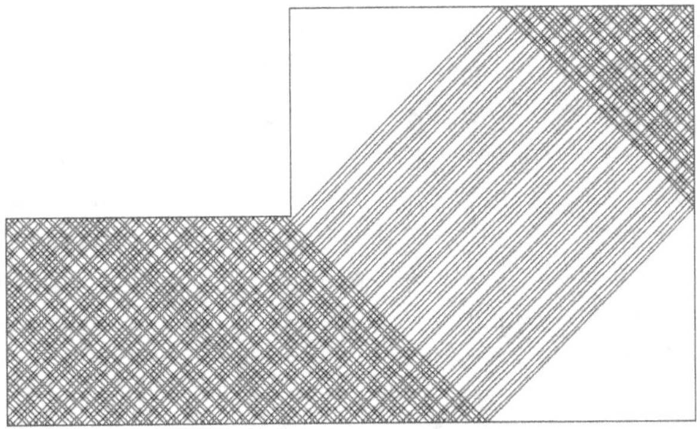

Fig. 3.78   A non-periodic trajectory that never enters the white corners.

gle and a portion of the square, and never visits the other two corners of the square. If we let the ball keep going, the shaded region would become completely black, but the corners would stay white.

The following question arises: which table shapes have the property that every non-periodic trajectory covers the whole table? It turns out that most of them do not. This is only true of tables that have a lot of symmetry, such as regular polygons, tables made from multiple squares glued together, and some simple triangles. For some examples of table shapes that have this property see [16], [117], [119] and previous sections of this Chapter.

---

[10]This question should be understood properly: Periodic trajectories appear in parallel families; we will count the number of such families.

3.7 Chaotic billiards

In theory of dynamical system a *chaotic* system is a dynamical systems, where the behavior of which highly sensitive to initial conditions.

Thus small differences in initial conditions, such as those due to rounding errors in numerical computation, yield widely diverging outcomes for such dynamical systems, rendering long-term prediction of their behavior impossible in general [67]. Consequently, a chaos is when the present determines the future, but the approximate present does not approximately determine the future.

Systems with chaotic behaviors exist in several disciplines, including meteorology, anthropology, sociology, physics, environmental science, computer science, engineering, economics, biology, ecology, and philosophy.

Note that [53] even simple processes can lead to chaos. That's why it's so hard to predict the weather, the stock market, and all sorts of other processes we come across in everyday life. One beautiful example is the game of billiards. As we have seen in the previous sections billiard trajectories may be periodic. But it turns out that this regular behavior is very rare.

It is known that for the vast majority of initial directions the trajectory will be much wilder: not only will it not retrace its steps, but it will eventually explore the whole of the table, getting arbitrarily close to every point on it. Moreover, a typical trajectory will visit each part of the table in equal measure, i.e., if consider two regions of the table whose areas are equal, then the trajectory will spend an equal amount of time in both. This behavior is a consequence of billiards being ergodic. The ergodicity of billiards means that it's really hard to predict where a ball will be after a given amount of time: in order to find out, one has to literally trace its path on piece of paper, meticulously measuring angle after angle, because one can not rely on any regular pattern to kick in.

If one gets the initial conditions of the ball (i.e., the place and direction it starts out with) ever so slightly wrong, then the little error will generally snowball, rendering the prediction inaccurate. This sensitive dependence on initial conditions is popularly known as the butterfly effect, and is one of the hall marks of chaos.

As far as real billiards goes, chaos is probably what makes it fun. In the 19th century the physicist L. Boltzmann suggested that many dynamical systems are ergodic: they are so chaotic that if one looks at the trajectory of a typical point this will in some sense explore all of the possibilities that are theoretically allowed. Well-known Birkhoff's ergodic theorem states that if a system is ergodic, then even though one cannot predict precisely what it will do in the future, one can accurately predict average quantities for typical trajectories. For example, in billiards, one might not be able to predict exactly where a ball will be after some time has elapsed, but one can accurately predict what proportion of its time it spends in a certain region of the table.

Remark 3.10. Note that the reader of the theory of chaotic billiards should be an advanced reader, familiar with Measure theory, Probability theory, Ergodic theory and dynamical systems. Moreover, the theory of chaotic dynamics is very large, this is why I am not going to give this theory here (this theory is too heavy for this book representing a simple introduction to the mathematical billiards). A reader interested to chaotic billiard can read [4], [8], [24], [25], [34], [81], [104], [108], [123], [124].

Bibliographical notes. This Chapter is the biggest chapter of the book. Because the theory of mathematical billiards is highly connected with geometry. There are many literatures mentioning geometrical properties of the table of mathematical billiards and their trajectories. In this chapter I have chosen the materials which (in my point of view) are simple and minimum for an introduction to the theory.

In the beginning part of each subsection I have citations to the source of the material. Here is the list of all bibliography used to write this chapter: [3], [5], [16], [20], [26]-[33], [37]-[41], [53]-[67], [74], [77], [79], [82], [86], [97]-[101], [106], [108]-[111], [117]-[125] and many internet sources.

Chapter 4

Billiards and physics

Mathematical billiards arise in many problems of physics. For example, one may consider billiards in potential fields. Another interesting modification, popular in the physical literature, is the billiard in a magnetic field [11], [110], [111]. Here we give some simple examples from physics which are related to billiards.

4.1 Phase space

Consider the movement of some system of objects - particles, bodies, etc. Let us add to space of configurations $\{x\}$ their speed vectors $\{v\}$), i.e. we will consider various couples $\{(x,v)\}$. The received set of couples (x is situation, v is speed) is called phase space of system, and the point (x,v) representing the system is called the phase point of the system. In other words, phase space is the set of various states of the movement of the system.

Example 4.1.

- Consider a point x on the real line $\mathbb{R}^1$ which can move on it with arbitrary speed v. Then the phase space will be $\mathbb{R}^2 = \{(x,v)\}$.
- The phase space of two independent points on $\mathbb{R}$ will be a 4-dimensional space.

A process is called determined if all its future states and all its past are defined by a state at the given moment. Such processes are, for example, processes of radioactive disintegration and reproduction of bacteria. There are also nondeterministic processes: such is the movement of particles in the quantum mechanics (neither the past, nor the future are defined by the present state) or distribution of heat (the future is defined by the present, and the past is not defined by the present).

The classical mechanics considers of the movement of systems whose future and the past are defined by initial states and speeds of all points of system, i.e. considers the determined processes. In each time point t_0 the mechanical system S is in the state $x(t_0)$ and has a concrete vector $v(t_0)$ of speed, a phase point of this system is $(x(t_0), v(t_0))$ which completely characterizes evolution of system as in the past, at all $t < t_0$, and as well as in the future, at all $t > t_0$. Knowing the one-unique state $S(0) = (x(t_0), v(t_0))$ of the movement of the system at the time t_0, it will be possible to define state of the movement of this system in all time is given by $\{S(t)\} = \{(x(t), v(t)), -\infty < t < \infty\}$. The set of states of the movement $\{S(t)\}$ is the set of all phase points $(x(t), v(t))$, which is called phase trajectory.

In phase space initial states of a system S can occupy some domain $U = U(0)$. At the time t all points of the domain U moving and generating the phase trajectories, appear in new points of phase space and form the new domain $U(t)$. How are $U(0)$ and $U(t)$ connected among themselves? It appears, in this respect for many mechanical systems (in particular, for so-called Hamilton systems) the following remarkable theorem is known.

Theorem 4.1. *(Liouville) Volumes of the domains $U(0)$ and $U(t)$ are identical.*

Proceeding from this theorem, all phase space can be imagined as the filled incompressible liquid which moves in the phase space (incompressibility and means preservation of volumes). Let us provide an example of the movement of such "phase liquid".

Example 4.2. We investigate the behavior of states of nine planets of the Solar System. For simplicity we will consider that every one of them represent a point mass 1 and moves around the Sun on its corresponding circular orbit without influencing other planets. Then the phase space of this system is the 54-dimensional space $\mathbb{R}^{54}$, since each planet has three space coordinates (allocate $3 \times 9 = 27$-dimensional configuration space $\mathbb{R}^{27}$) and three coordinates of the vector of speed (allocate $3 \times 9 = 27$-dimensional space of speeds). However as each planet with number k (planets are numbered in order of the distance from the Sun) moves on the circle of its orbit, its space coordinates $(x_{1,k}, x_{2,k}, x_{3,k})$, denoted by R_k, the orbit radius of the k-th planet will have a configuration point that satisfies $x_{1,k}^2 + x_{2,k}^2 + x_{3,k}^2 = R_k^2$,

$k = 1, 2, \ldots, 9$. A configuration point x represents the position of all nine planets, does not move on all 27-dimensional space, but only on its limited part – the so-called 9-dimensional torus $\mathbb{T}^9 = \underbrace{S^1 \times \cdots \times S^1}_{9 \ \text{times}}$. Moreover the absolute value of each speed vector is constant:

$$|v_k| = \sqrt{v_{1,k}^2 + v_{2,k}^2 + v_{3,k}^2} = const, \quad k = 1, 2, \ldots, 9.$$

Each of these nine equalities sets the two-dimensional sphere, and all equalities together separate in 27-dimensional space of speeds a limited part – 18-dimensional manifold $\mathrm{M}^{18} = \underbrace{S^2 \times \cdots \times S^2}_{9 \ \text{times}}$. Thus in the phase space $\mathbb{R}^{54}$ one considers $9 + 18 = 27$-dimensional set $D = \mathbb{T}^9 \times \mathrm{M}^{18}$ which transfers to itself by the phase flow map T: $T(D) = D$. If $(x(0), v(0)) \in D$ is a phase point of the Solar System at moment $t = 0$ and U is its arbitrary neighborhood, then U is a domain in D, volume of which under the mapping T is invariant: $volume(T(U)) = volume(U)$ (by Liouville's theorem). Hence we are at conditions of the Poincaré recurrence Theorem 1.6 as it follows from this theorem that after some time all planets will be near the initial positions and will have approximately initial speeds: at some n the point $T^n(x(0), v(0))$ will lie in the same neighborhood U, as the initial phase point $(x(0), v(0))$.

Exercise 37. Around the Sun a planet with a radius R of orbit rotates and around it (in the same plane) a satellite with orbit radius r rotates, and angular speeds of their rotation are not commensurable. Prove that the satellite will fill everywhere densely the ring with the center in the Sun and with internal radius $R - r$, and external radius $R + r$.

4.2   Physics of billiards

Mathematical billiards are much simplified version of real game of billiard. The following subsection provides physical properties of the usual billiard game.

4.2.1   *Motion and collisions of balls*

Here following an internet article[1] we give the physics behind billiards, in large part, involves collisions between billiard balls. Assume that when

---

[1] https://www.real-world-physics-problems.com/physics-of-billiards.html

two billiard balls collide the collision is nearly elastic. An elastic collision is one in which the kinetic energy of the system is conserved before and after impact. Therefore, for simplicity one can assume that for collisions involving billiard balls, the collision is perfectly elastic.

It is also assumed that for collisions between balls, momentum is always conserved and is frictionless. By elastic-collision assumption we can find the trajectory of two colliding billiard balls. The Fig. 4.1 shows a collision between two billiard balls.

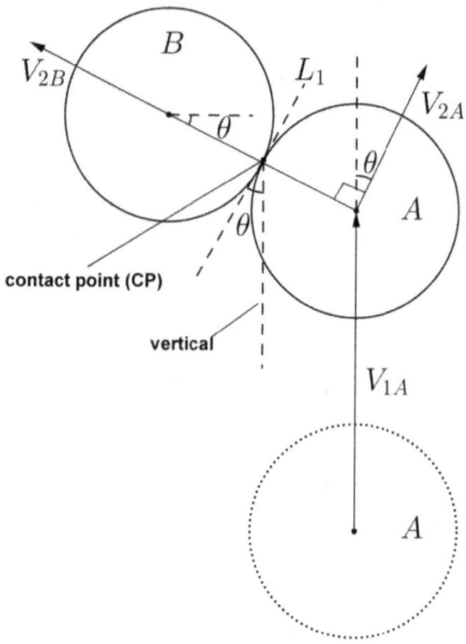

Fig. 4.1

Consider two balls A and B having the same mass. Assume that ball B initially has zero velocity (at rest). The initial velocity of ball A is V_{1A}. After impact, ball A moves at velocity V_{2A} in the direction shown in Fig. 4.1, and ball B moves at velocity V_{2B} in the direction shown. Denote by L_1 the line which is tangent to both balls at the point of contact. Thus L_1 is perpendicular to the line passing through the center of the two balls and the contact point (CP). It is clear that if L_1 makes an angle θ with the vertical, then the line passing through the center of the balls makes the angle θ with the horizontal.

After impact at CP, ball B moves in the direction of the line joining the center of the two balls (see Fig. 4.1). This is because the impulse force delivered by ball A to ball B acts normal to the surface of ball B, assuming there is no friction between the balls. Consequently, ball B moves in the direction of this impulse.

Proposition 4.1. *After impact, ball A moves in a direction perpendicular to the direction of ball B.*

Proof. For the two colliding balls, the general vector equation for conservation of linear momentum is:

$$m_A \vec{V}_{1A} = m_A \vec{V}_{2A} + m_B \vec{V}_{2B}.$$

Since the masses are equal: $m_A = m_B$, from the last equation we get

$$\vec{V}_{1A} = \vec{V}_{2A} + \vec{V}_{2B}.$$

For an elastic collision kinetic energy is conserved, and the equation is:

$$\frac{1}{2} m_A V_{1A}^2 = \frac{1}{2} m_A V_{2A}^2 + \frac{1}{2} m_B V_{2B}^2.$$

Consequently, again using $m_A = m_B$ we get

$$V_{1A}^2 = V_{2A}^2 + V_{2B}^2.$$

By the Pythagorean theorem, the last equation tells us that the vectors $\vec{V}_{1A}, \vec{V}_{2A}, \vec{V}_{2B}$ form a right triangle. Moreover, $\vec{V}_{2A} \perp \vec{V}_{2B}$. Thus, after impact ball A moves in a direction perpendicular to the direction of ball B. $\square$

Consider two special cases involving ball collision:

The case where the target ball B must be hit at an angle θ very close to zero, ball A needs to be moving at a high speed V_{1A}, meaning one would have to hit ball A quite hard with the cue. Because only a very small fraction of the momentum of ball A (and therefore velocity) is transferred to ball B, due to the obliqueness of the impact.

The case where the impact is head on ($\theta = 90°$) the above solution does not apply. In this case $V_{2A} = 0$ and $V_{2B} = V_{1A}$. This essentially means that the velocity of ball A is completely transferred to ball B.

It is known that in the physics of hitting a baseball there is a sweet spot on a ball, similarly in physics of billiards there is such a sweet spot where one can strike with the cue stick so that no friction force develops between the ball and the billiard table. Knowing where this sweet spot can give an idea of where to hit the ball so that it develops back or forward spin, which can be useful when making a shot.

Take the position of the cue (the force F) at height h see Fig. 4.2. We are going to find the height h so that no (horizontal) frictional force develops at point P when the ball is struck by the cue.

We represent the ball $+$ and cue system with a free-body diagram as shown in Fig. 4.2. In Fig. 4.2 we have the following notations:

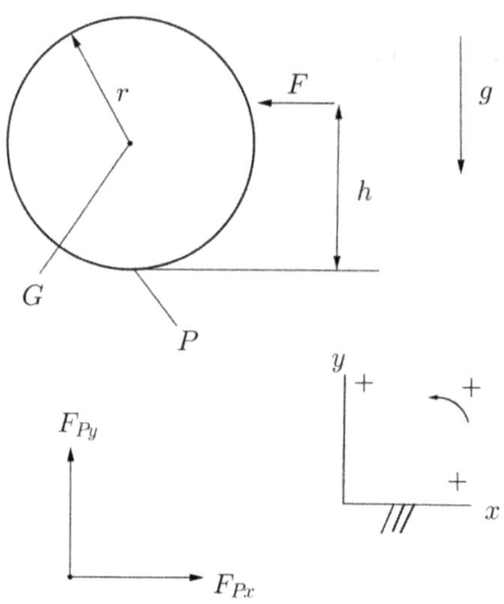

Fig. 4.2

F is the force the cue exerts on the ball when it strikes

r is the radius of the ball

G is the center of mass of the ball

g is the acceleration due to gravity, which is 9.8 m/s^2

P is the point of contact of the ball with the billiard table

F_{P_x} is the x-component of the force exerted on the ball by the billiard table, at point P. This is a frictional force.

F_{P_y} is the y-component of the force exerted on the ball by the billiard table, at point P.

By Newton's Second Law, the general force equation in the x-direction is:

$$\sum F_x = ma_{Gx},$$

where m is the mass of the ball a_{Gx} is the acceleration of the center of mass in the x-direction. For our system this equation becomes

$$F_{P_x} - F = ma_{Gx}.$$

Since $F_{P_x} = 0$ we get

$$-F = ma_{Gx}. \tag{4.1}$$

Similarly, the general force equation in the y-direction is:

$$\sum F_y = ma_{Gy},$$

where a_{Gy} is the acceleration of the center of mass in the y-direction.

Since the billiard ball only moves in the x-direction $a_{G_y} = 0$, the above equation becomes

$$F_{P_y} - mg = a_{G_y} = 0,$$

Consequently,

$$F_{P_y} = mg.$$

Now we write the general moment equation for rotation of a rigid body about its center of mass G.

$$\sum M_G = I_G \alpha,$$

where $\sum M_G$ is the sum of the moments about the center of mass G, I_G is the moment of inertia of the ball about its center of mass and α is the angular acceleration of the ball.

Since no frictional force develops between the ball and table, there is no relative slipping at point P. This means that we have a case of pure rolling. Consequently we can write

$$\alpha = -\frac{a_{Gx}}{r},$$

where the negative sign is there to match the sign convention used in this problem.

Thus the moment equation becomes

$$F \cdot (h - r) = I_G \left(-\frac{a_{Gx}}{r} \right). \tag{4.2}$$

By equations (4.1) and (4.2) we get

$$h = \frac{I_G}{mr} + r.$$

For a solid sphere

$$I_G = \frac{2}{5} mr^2.$$

Consequently,

$$h = \frac{7r}{5}.$$

Thus this is the height to hit the ball so that no friction develops at point P. No matter how hard one hits the ball at this location, no friction (reaction) force will develop at point P. Therefore, pure rolling of the ball will always result after impact (no relative slipping).

Moreover, in the cases where the cue strikes above or below this height h, friction is necessary to prevent the ball from slipping on the surface of the billiard table. If the ball is hit hard enough, above or below height h, relative slipping will occur, due to insufficient friction between ball and table.

Consider two balls, with masses m_1, m_2 and moving on an finite segment. Let v_1 and v_2 be their speeds at the initial moment, then their speeds u_1 and u_2 after a collision can be computed by the equations (conversation law of impulse and energy):

$$m_1 u_1 + m_2 u_2 = m_1 v_1 + m_2 v_2, \tag{4.3}$$

$$m_1 u_1^2 + m_2 u_2^2 = m_1 v_1^2 + m_2 v_2^2. \tag{4.4}$$

Denote by $\Delta I_i = m_i(u_i - v_i)$ the difference of the impulses of the i-th ball.

Exercise 38. Using (4.3) and (4.4) prove the following formulas

$$u_1 = \frac{1}{m_1 + m_2}[(m_1 - m_2)v_1 + 2m_2 v_2],$$

$$u_2 = \frac{1}{m_1 + m_2}[2m_1 v_1 + (m_2 - m_1)v_2],$$

$$\Delta I_1 = \frac{m_1 m_2}{m_1 + m_2}(v_2 - v_1), \quad \Delta I_2 = \frac{m_1 m_2}{m_1 + m_2}(v_1 - v_2).$$

Exercise 39. Prove that after the collision of balls of equal mass they exchange speeds. That is $u_1 = v_2$, $u_2 = v_1$.

4.2.2 *Fermat principle*

Let us give an example from geometrical optics. According to the Fermat principle, light propagates from point A to point B in the least possible time. In Euclidean geometry, that is a homogeneous and isotropic medium, the Fermat principle means that light "chooses" the straight line AB. Now

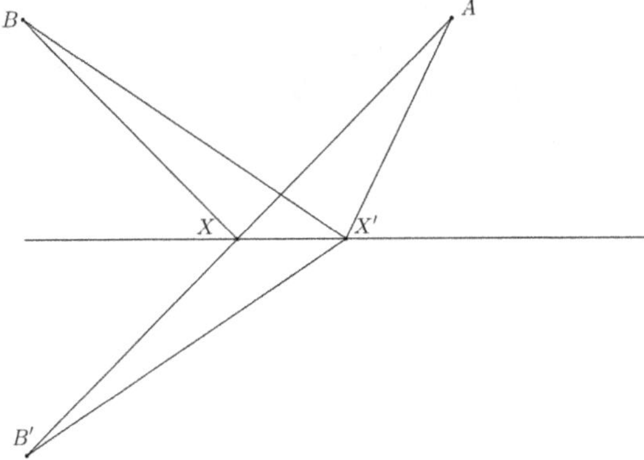

Fig. 4.3   The reflection in a straight line mirror

we consider a single reflection in a mirror that is assumed to be a straight line l in the plane (Fig. 4.3).

We are interested to find a broken line AXB (where $X \in l$) of minimal length. To find the position of point X, reflect point B in the mirror and connect to A. It is easy to see that for any other position X' of point X, the broken line $AX'B$ is longer than AXB. This construction implies that the angles made by the incoming and outgoing rays AX and XB with the mirror l are equal. Therefore, we obtain the billiard reflection law as a consequence of the Fermat principle.

Let the mirror be an arbitrary smooth curve l (Fig. 4.4). Take points $A = (a_1, a_2)$ and $B = (b_1, b_2)$. In this case the reflection point $X = (x_1, x_2)$ also extremizes the length of the broken line AXB. Now we are going to deduce the reflection law. Let X be a point of the plane, and define the function $f(X) = |AX| + |BX|$ as sum of distances $|AX|$, $|BX|$. The gradient of the function $|AX| = \sqrt{(x_1 - a_1)^2 + (x_2 - a_2)^2}$, i.e.,

$$\nabla |AX| = \left(\frac{x_1 - a_1}{\sqrt{(x_1 - a_1)^2 + (x_2 - a_2)^2}}, \; \frac{x_2 - a_2}{\sqrt{(x_1 - a_1)^2 + (x_2 - a_2)^2}} \right)$$

is the unit vector in the direction from A to X, i.e., the vector $AX = (x_1 - a_1, x_2 - a_2)$ and similarly for $|BX|$. We are interested in critical points of $f(X)$, subject to the constraint $X \in l$. By the Lagrange multipliers principle (see for example[2]), X is a critical point if and only if $\nabla f(X) =$

---

[2]https://en.wikipedia.org/wiki/Lagrange_multiplier

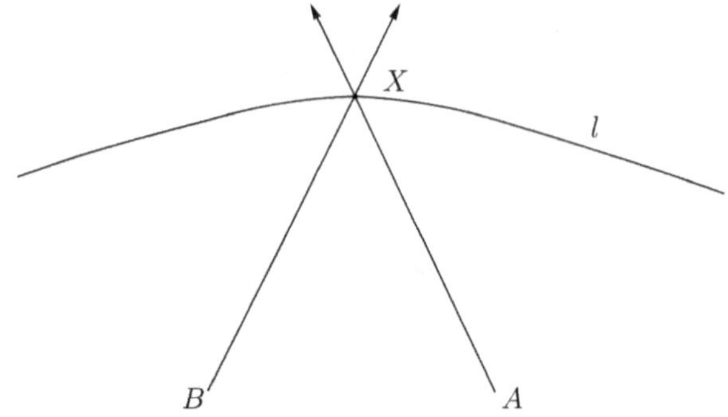

Fig. 4.4   The reflection in a smooth curved mirror

$\bigtriangledown|AX|+\bigtriangledown|BX|$ is orthogonal to l. The sum of the unit vectors from A to X and from B to X is perpendicular to l if and only if AX and BX make equal angles with l. We have again obtained the billiard reflection law. Of course, the same argument works if the mirror is a smooth hypersurface in multi-dimensional space, and in Riemannian geometries other than Euclidean.

The above argument could be rephrased using a different mechanical model. Let l be wire, X a small ring that can move along the wire without friction, and AXB an elastic string fixed at points A and B. The string assumes minimal length, and the equilibrium condition for the ring X is that the sum of the two equal tension forces along the segments XA and XB is orthogonal to l. This implies the equal angles condition.

4.3   Mechanical interpretations of three-periodic points

Following [39] we give a mechanical interpretation of the 3-periodic trajectory (see subsection 3.3.1) in an acute triangle. Assume that on each side of $\triangle ABC$ a short wire and a ringlet is laid, where these ringlets may move freely along the sides. The neighboring ringlets are connected by a rubber band and are set free. If the stationary state is reached, the ringlets and rubber bands will form the vertices and sides of a triangle $H_1H_2H_3$ of least perimeter and inscribed in $\triangle ABC$. Moreover, the equilibrium forces acting on neighboring ringlets are orthogonal to the sides of $\triangle ABC$ and are directed along its altitudes. This follows from the fact that the tensions of all ringlets in the stationary state are equal to one another. Thus, $H_1H_2H_3$ is a periodic billiard trajectory, see Fig. 4.5.

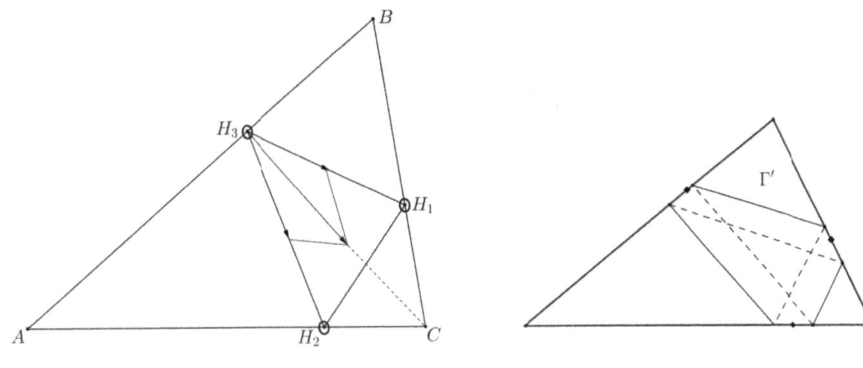

Fig. 4.5

Now let us give another mechanical interpretation of three periodic trajectory in a triangle: If the triangle ABC is considered to be a plane smooth plate, then the twice-traversed three periodic trajectory Γ and each 6-periodic trajectory Γ' in the pencil of trajectories parallel to Γ can be represented as an elastic closed thread winding around this plate and in turn passing from one side to another (Fig. 4.5).

Another example is the Sinai's billiard: its boundary ∂Q consists of finitely many smooth arcs, convex inside Q (see Fig. 4.6).

For this billiard there is a periodic trajectory. Sinai's billiard has the same mechanical interpretation of periodic trajectories as for an acute triangle. Namely, on the boundary components one puts ringlets and connect them by elastic bands, and we let this construction move freely. The bands tighten, and form a periodic billiard trajectory. As in the case of an acute triangle, in Sinai's billiards the length of each periodic trajectory is minimal among the lengths of all polygonal lines with vertices on the same boundary sides as the reflection points of the trajectory; moreover, this minimum is strict, that is, unique.

4.4   Billiard trajectories of light

The following are reformulation of properties of ellipse and parabola which we proved in Chapter 3:

Proposition 4.2. *A ray of light through a focus of an ellipse reflects to a ray that passes through the other focus. A ray of light through a focus of a parabola reflects to a ray parallel to the axis of the parabola.*

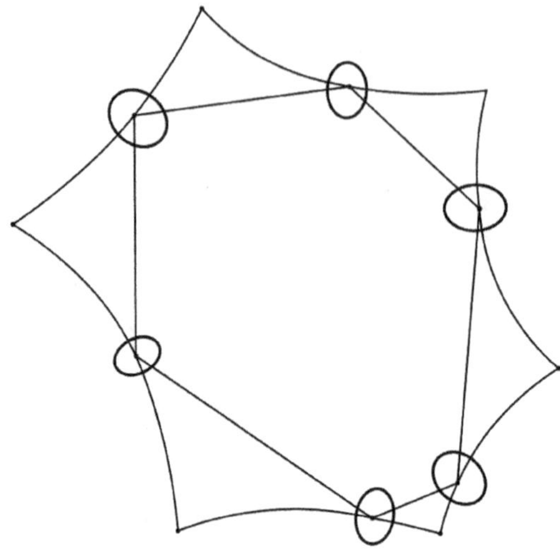

Fig. 4.6

There are some potential applications to the design of optical cavities for lasers.

Imagine a region whose sides are mirrored. One can shine a laser in, and have an opening or a partial reflector where the light can come out. Typically, there is some "gain region" in the interior, say a crystal which is excited electrically or by a laser operating at another frequency. We want some input beam to pass through the gain region many times, or to spend a lot of time there on average, before the beam hits the output.

A complication not present in the usual mathematical billiards is that the gain region may have a different index of refraction from the surrounding medium, so light entering it at an angle may be deflected. This can even depend on the intensity of the light.

One possibility is to control the geometry very precisely. In fact, this is one place one can use the fact that hyperboloids of one sheet are doubly ruled surfaces: make the even segments follow one ruling while the odd segments follow the other ruling. However, aligning this precisely can be tricky, particularly with the complications above, and sometimes it does not make efficient use of the volume, so one does get as many passes through the crystal for the space you allocate to the cavity.

There are applications of optical properties of ellipse [2], [37], [87], [110]. Here we give some of them:

4.4.1 *A construction of a trap for light*

A construction of a trap for a beam of light [110], that is, a reflecting curve such that parallel rays of light, shone into it, get permanently trapped. One of such constructions is given in [87] (see Fig. 4.7) Here the curve γ is a

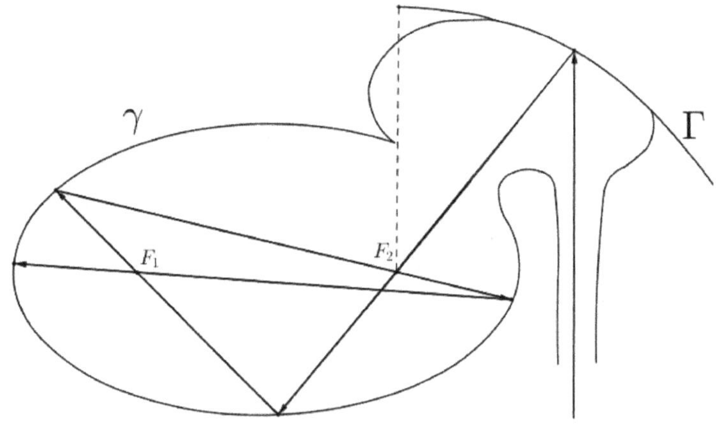

Fig. 4.7   A trap for a beam of light

part of an ellipse with foci F_1 and F_2 the curve Γ is a parabola with focus F_2. These curves are joined in a smooth way to produce a trap: it follows from Section 3.4 that a vertical ray, entering the curve through a window, will tend to the major axis of the ellipse and will therefore never escape.

The following proposition about a revisit of trap for a parallel beam of light.

Proposition 4.3. *A set U of rays of light, having a positive area, cannot be trapped.*

Proof.   Assume that such a trap exists. Close the entrance window by a reflecting curve δ to obtain a billiard table. The phase space of this billiard has a finite area, and the billiard ball transformation T is area preserving. Consider the incoming rays from the set U as phase points with foot points on δ. By Poincaré's recurrence Theorem 1.6, there exists a phase point in

U whose T^n trajectory returns to U. This means that the respective ray of light will eventually hit δ and escape from the trap, a contradiction. $\square$

4.4.2 *Corner reflector*

A reflector is a device or surface that reflects light back to its source with a minimum of scattering. In a reflector an electromagnetic wavefront is reflected back along a vector that is parallel to but opposite in direction from the wave's source (see[3] and [2] page 26.)

Corner reflector is a set of three mutually perpendicular reflective surfaces, placed to form the corner of a cube, work as a reflector. The three corresponding normal vectors of the corner's sides form a basis (x, y, z) in which to represent the direction of an arbitrary incoming ray, $[a, b, c]$. When the ray reflects from the first side, say x, the ray's x-component, a, is reversed to $-a$, while the y- and z-components are unchanged. Therefore, as the ray reflects first from side x then side y and finally from side z the ray direction goes from $[a, b, c]$ to $[-a, b, c]$ to $[-a, -b, c]$ to $[-a, -b, -c]$ and it leaves the corner with all three components of its direction exactly reversed (Fig. 4.8) Corner reflectors occur in two varieties. In the more common

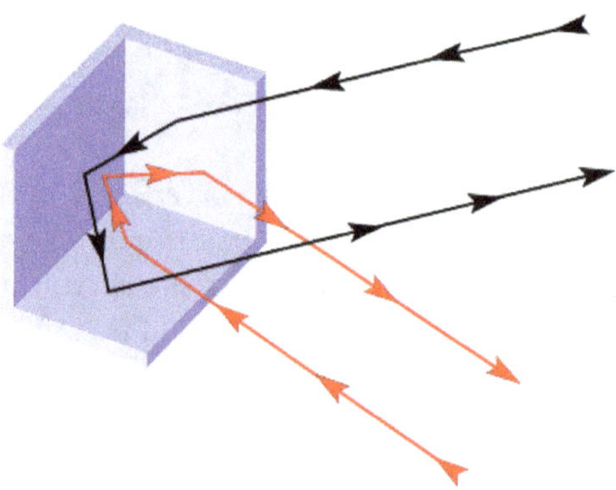

Fig. 4.8   Working principle of a corner reflector

form, the corner is literally the truncated corner of a cube of transparent material such as conventional optical glass. In this structure, the reflec-

---

[3]https://en.wikipedia.org/wiki/Retroreflector

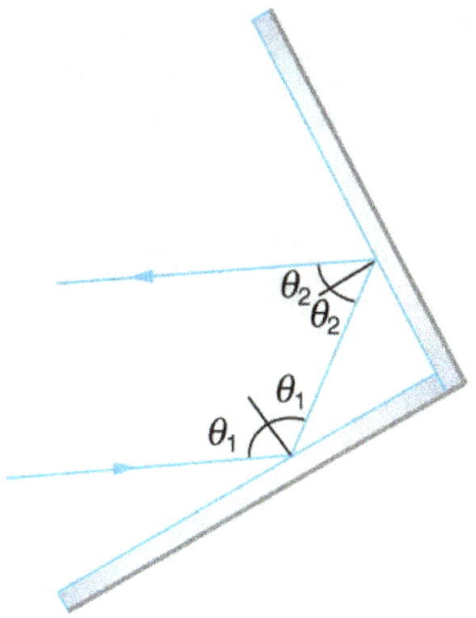

Fig. 4.9   A corner reflector sends the reflected ray back in a direction parallel to the incident ray, independent of incoming direction.

tion is achieved either by total internal reflection or silvering of the outer cube surfaces. The second form uses mutually perpendicular flat mirrors bracketing an air space. These two types have similar optical properties.

A large relatively thin reflector can be formed by combining many small corner reflectors, using the standard hexagonal tiling.

Exercise 40. Show that when light reflects from two mirrors that meet each other at a right angle, the outgoing ray is parallel to the incoming ray (see Fig. 4.9)[4].

4.4.3   *Crushing of stones in a kidney*

Lithotripsy is a medical procedure involving the physical destruction of hardened masses like kidney stones or gallstones (see[5] and [2] page 40). A lithotripter a machine that pulverizes kidney stones by ultrasound (stock waves) as an alternative to their surgical removal. The patient is placed in an elliptical tub with the kidney stone at one focus of the ellipse, see Fig. 4.10. A beam is projected from the other focus of the tub, so that it reflects to hit the kidney stone (see Theorem 3.15). Thus the shock waves

---

[4]https://courses.lumenlearning.com/austincc-physics2/chapter/25-2-the-law-of-reflection/

[5]https://www.kidney.org/atoz/content/kidneystones_shockwave

from outside the body are targeted at a kidney stone causing the stone to fragment. Then the stones are broken into tiny pieces.

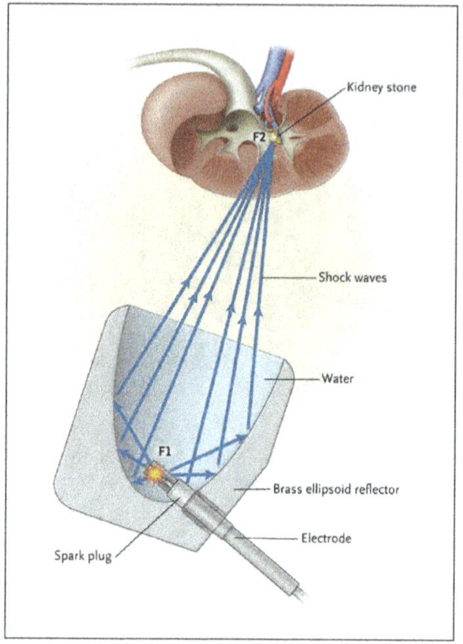

Fig. 4.10   Lithotripter for crushing a stone in kidney. Source: Internet.

Exercise 41. Assume a lithotripter is based on the ellipse with equation

$$\frac{x^2}{16} + \frac{y^2}{9} = 1.$$

How far from the center of the ellipse must the kidney stone and the source of the beam be placed?

Hint. See (3.4) and use that $c^2 = a^2 - b^2$, $a > b$.

4.4.4   *Lighting problems of a non-convex area*

The geometrical properties of an ellipse (Section 3.4), can be used to answer the following question: with what smallest number of light sources points (bulbs) it will be possible to light an interior of limited domain (area) Ω on the plane? [37].

The problem was first solved by R.Penrose (1958) using ellipses to form the Penrose unilluminable room.[6] He showed there exists a room with curved walls that must always have dark regions if lit only by a single point source (see Fig. 4.11).

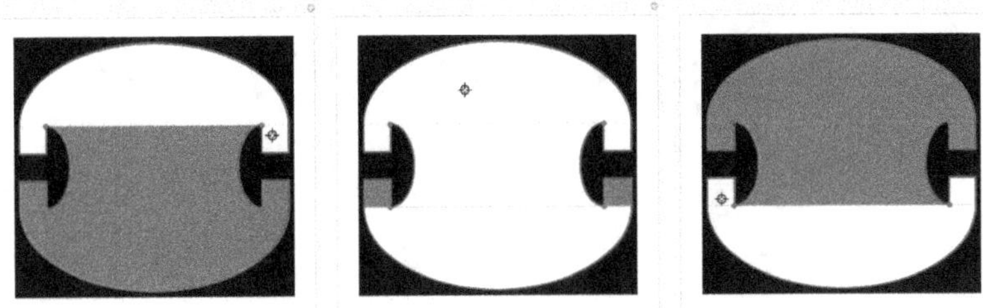

Fig. 4.11   Penrose's room which always has dark regions, regardless of the position of the bulb. Here the locator is the bulb, and regions of the room are white (lit) or gray (dark). The red points are the focuses of the half ellipses at the top and bottom of the room.

Exercise 42. Show the dark regions of Penrose's room when the bulb is inside of the rectangle with vertices on red points (i.e. focuses of the half ellipses).

This problem was also solved for polygonal rooms by G. Tokarsky (1995) for 2 dimensions, which showed there exists an unilluminable polygonal 26-sided room with a "dark spot" which is not illuminated from another point in the room, even allowing for repeated reflections [112]. This was a borderline case, however, since a finite number of dark points (rather than regions) are unilluminable from any given position of the point source. An improved solution was given D. Castro (1997), with a 24-sided room with the same properties.

The boundary $\partial\Omega$ of the area Ω is a closed not self-crossing curve γ on the plane thus, moving on the curve γ, we will leave domain Ω on one side from γ all the time.

It is clear that if the domain Ω is convex, then one bulb which can be placed in any place is sufficient for fully lighting. But if the domain Ω is not convex, then one bulb for lighting of the area may not be enough any more (see Fig. 4.12) the bulb O does not light any point A of the shaded area lying in. However, in Fig. 4.12 we just unsuccessfully chose location for a

---

[6]https://en.wikipedia.org/wiki/Illumination_problem

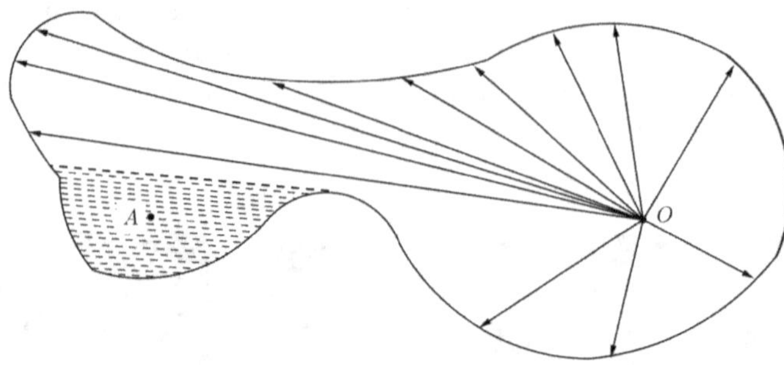

Fig. 4.12   The place of the bulb is not sufficient for fully lighting.

bulb: the point in Fig. 4.13 is already chosen more successfully, and the bulb O lights all area Ω. If a domain Ω contains a point O, from which all its other points "are visible" (as, for example, Fig. 4.13) then this domain Ω is called star-shaped, and the point O is called a visibility point.

We note that all convex sets belong to star-shaped sets, in particular. It is clear, that any star-shaped domain can be lit with one bulb, putting the bulb on the visible point. Let us assume now that Ω is not a star-shaped domain: its any point O is not "a visibility point". Then one bulb in what place of the domain we would not place it, completely will not light it, that is, illumination of Ω requires more than one bulb.

It is possible to assume that for lighting there will be needed a finite number of bulbs, independently on the kind of the domain Ω if only its boundary $\gamma = \partial\Omega$ has no break points, i.e., is rather smooth. It is confirmed by the strict theorem which we will provide in the end of this subsection. We already know that the movement of a ray of light at reflection from the

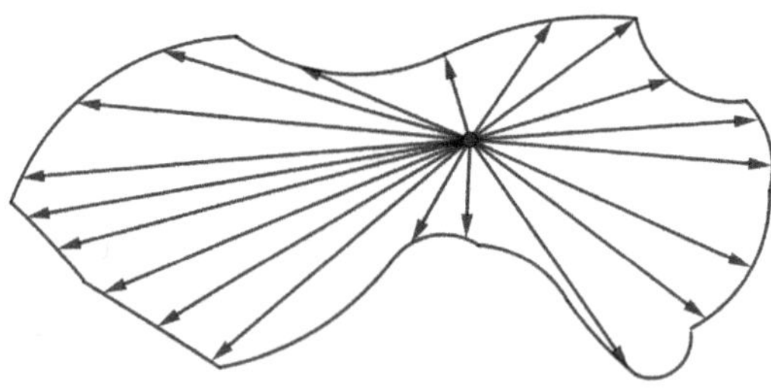

Fig. 4.13   One bulb is sufficient for lighting.

boundary of domain is carried out under the billiard law (law of equal angle reflections). Therefore the above given question has new sounding, we will consider the domain Ω with a mirror boundary, and any ray of light at hit point on boundary is reflected from it under the billiard's law.

Beginning from this moment and until the end of the paragraph we assume that Ω has a mirror boundary. In many such domains (for example, polygons) for lighting of its area one bulb suffices. But this property is not true in general. The following statements make this point clear [37]:

Proposition 4.4. *For any natural number n there exists a domain, with smooth boundary, for lighting of which it is necessary more than n bulbs (see Fig. 4.14).*

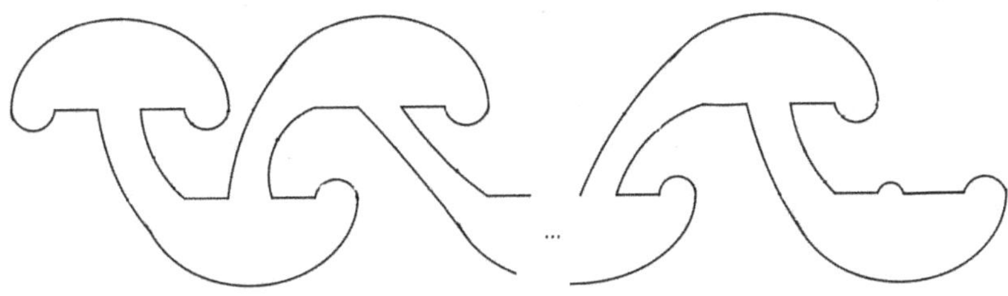

Fig. 4.14   The domain mentioned in Proposition 4.4.

Proposition 4.5. *There exists a bounded domain for lighting of which it is necessary infinitely many bulbs (see Fig. 4.15).*

Fig. 4.15   The domain mentioned in Proposition 4.5. The right end of the figure is geometrically degreasing infinite sequence of semi-ellipses.

The following theorem answers the main question of this subsection.

Theorem 4.2. *The domain Ω with the mirror boundary curve γ without breaks[7], can be lit by a finite number of bulbs if the curve γ is smooth (i.e. from the class C^1). Moreover, if the curve γ is not smooth at least on one point then a finite number of bulbs for lighting of the domain Ω may not be sufficient.*

4.5 The mechanical interpretation of billiard trajectories in right triangles

In this section, following [41], we formulate in terms of mechanics some results concerning billiard trajectories in right triangles.

Consider two elastic masses moving on the segment $[0, 1]$ and colliding with each other and with the ends of the segment. Denote the masses by m_1 and m_2, their initial positions by x_1 and x_2, and their initial velocities by v_1 and v_2. Suppose that the second particle is situated further to the right than the first one, that is $0 \le x_1 \le x_2 \le 1$.

Theorem 4.3.

(i) *Let $\alpha = \arctan \sqrt{\frac{m_1}{m_2}}$ be commensurable with π. Then the periodic orbits in such a system are everywhere dense in the phase space. Moreover, for given x_1, x_2, v_1, and v_2, an arbitrarily small ϵ, and an arbitrarily great integer N, it is possible to change $x_1, x_2, v_1,$ and v_2 by less than ϵ so that the motion of the points be periodic and the period be greater than N^8.*

(ii) *For any m_1, m_2, x_1, x_2, and $v_1 \ne 0$ there exists a v_2 such that the motion is periodic.*

(iii) *Suppose $x_1 = 0$, $m_1 > 3m_2$. Then for any x_2 and $v_1 \ne 0$ there exists a v_2 such that the motion is periodic. Moreover, the period can be made as big as needed.*

(iv) *For arbitrary masses m_1, m_2 and $x_1 = 0$, $v_2 = 0$, $v_1 \ne 0$ for almost all x_2 the motion of the system is periodic.*

Proof. The proof can be realized by considering the configuration space of the system, i.e., the right triangle with acute angle α [103]. First, we will explain why the movement of the masses is equivalent to the movement of a particle in the right triangle with acute angle α. Put $y_1 = \sqrt{m_1}\, x_1$,

---

[7]At each point of γ it is possible to carry out a unique tangent line
[8]The number of bounces in the period is greater than N

$y_2 = \sqrt{m_2}\, x_2$. Let us represent the position of the masses by the point with coordinates (y_1, y_2) in the plane. This point will play the role of a billiard particle (a ball). Its velocity vector $\vec{u}$ equals $\vec{u} = (u_1, u_2) = (\sqrt{m_1}v_1, \sqrt{m_2}v_2)$. The configuration space of the system is determined by the conditions

$$0 \leq x_1 \leq 1 \quad \Leftrightarrow \quad 0 \leq y_1 \leq \sqrt{m_1},$$

$$0 \leq x_2 \leq 1 \quad \Leftrightarrow \quad 0 \leq y_2 \leq \sqrt{m_2},$$

$$x_1 \leq x_2 \quad \Leftrightarrow \quad \sqrt{m_1}y_1 \leq \sqrt{m_2}y_2.$$

Thus, the configuration space is the right triangle with legs equal to $\sqrt{m_1}$ and $\sqrt{m_2}$ (see Fig. 4.16).

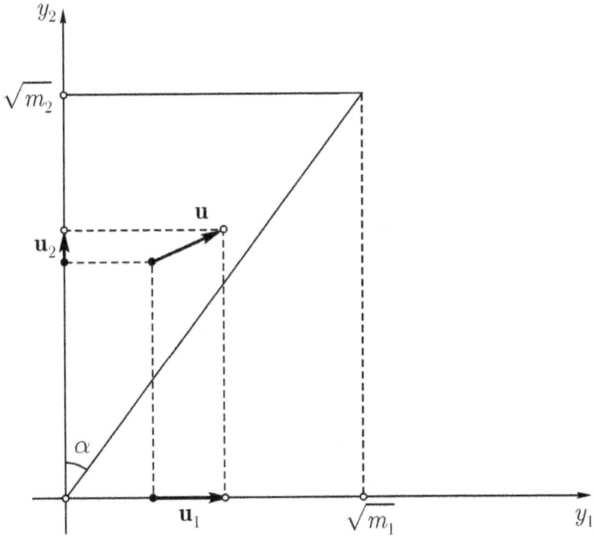

Fig. 4.16

A collision of a mass with an extremity of the segment corresponds to a collision of the billiard particle with a leg of the triangle, while a collision between the masses corresponds to a collision of the particle with the hypotenuse. What we have to show is that the collisions of the particle obey the reflection law. This is almost obvious for a collision with a leg. Indeed, suppose that, say, the first mass collides with the left extremity of the segment. Then v_2 (and therefore u_2) do not change, while v_1 (and therefore u_1) change sign. This means that the particle is reflected in the left leg of the triangle accordingly to the reflection law. The case of the other leg is similar. Now consider a collision between the masses. We denote by v_1', v_2', u_1', u_2', and $\vec{u}'$ the values of v_1, v_2, u_1, u_2, and $\vec{u}$ after the collision. Moreover, we denote by $\vec{h}$ the vector represented by hypotenuse of the triangle, i.e., $\vec{h} = (\sqrt{m_1}, \sqrt{m_2})$. The law of conservation of energy gives

$$m_1 v_1^2 + m_2 v_2^2 = m_1 v'^2_1 + m_2 v'^2_2 \quad \Leftrightarrow$$

$$(\sqrt{m_1} v_1)^2 + (\sqrt{m_2} v_2)^2 = (\sqrt{m_1} v'_1)^2 + (\sqrt{m_2} v'_2)^2 \quad \Leftrightarrow$$

$$\|\vec{u}\| = \|\vec{u}'\|.$$

The law of conservation of momentum gives

$$m_1 v_1 + m_2 v_2 = m_1 v_1' + m_2 v_2',$$

which is equivalent to

$$\sqrt{m_1} \cdot (\sqrt{m_1} v_1) + \sqrt{m_2} \cdot (\sqrt{m_2} v_2) = \sqrt{m_1} \cdot (\sqrt{m_1} v_1') + \sqrt{m_2} \cdot (\sqrt{m_2} v_2').$$

That is

$$\vec{h} \cdot \vec{u} = \vec{h} \cdot \vec{u}'.$$

Thus after the collision, the velocity vector $\vec{u}$ conserves both its norm and the angle that it makes with the hypotenuse of the triangle. This means that the collision obeys the reflection law. Hence we have established the equivalence between the mechanical system of two masses on a segment and the billiards in the right triangle with acute angles equal to $\alpha = \arctan \sqrt{\frac{m_1}{m_2}}$ and $\beta = \arctan \sqrt{\frac{m_2}{m_1}}$. Using this equivalence we give the proof of the theorem:

(i) The first part of the theorem follows from the fact [14]: for a billiard particle (a ball) in a rational triangle and for given ϵ, N and for any initial position and velocity of the particle, the position and the velocity can be changed by less than ϵ in such a way that the trajectory of the particle becomes periodic and the period is greater than N.

(ii) The second part of the theorem follows from Theorem 3.6, which states that the billiard particle can be launched from every point of a right triangle so that the trajectory will be periodic. Moreover, it follows from the construction that the particle can be launched in a direction not parallel to the legs of the triangle. Therefore it suffices to take v_2 such that the slope of the trajectory at the point is equal to the slope of $\vec{u}$ in Fig. 4.16, i.e.,

$$\frac{u_2}{u_1} = \frac{\sqrt{m_2}v_2}{\sqrt{m_1}v_1} = \text{slope of the trajectory.}$$

v_2 always exists, because $v_1 \neq 0$.

(iii) To prove the third part it suffices to consider the mirror trajectories (see Subsection 3.3.4) of launched from the smaller leg. Since $m_1 > 3m_2$, the smaller acute angle α of the triangle satisfies

$$\alpha = \arctan\sqrt{\frac{m_1}{m_2}} \leq \arctan\left(\frac{1}{\sqrt{3}}\right) = \frac{\pi}{6}.$$

Consequently there are infinitely many types of mirror periodic trajectories and the period can be made as big as needed.

(iv) The motion of the system corresponds to perpendicular trajectories. According to Subsection 3.3.2 almost all such trajectories and therefore motions of the system are periodic, that is, the set of points on the sides of a right triangle from which non-periodic (perpendicular) trajectories begin is a set of measure zero. $\square$

Exercise 43. If two balls of equal mass, $m_1 = m_2$, move on a segment OA so that their initial speeds v_1 and v_2 are incommensurable, then for any two small intervals of OA there exists a moment t of time when the first ball will be in the first interval, and the second ball will be in the second interval.

4.6 Billiard of elementary one-dimensional elastic collisions of three particles

This section is devoted to billiard theory of elementary one-dimensional elastic collisions of balls, we follow [89]. Here we give a simple connection between the motion of few-particle elastically colliding systems in one dimension and a corresponding billiard system.

Note that the collision behaviors of the or two-particle (two-ball) and barrier system were given in Sections 2.5.4 and 4.5. We now consider three elastically colliding particles of arbitrary masses on an infinite one-dimensional line. This system can be mapped onto the motion of a billiard ball in an infinite wedge whose opening angle depend on the three masses. Moreover, we discuss the motion of three particles on a finite ring. This system can be mapped onto the motion of a billiard ball on a triangular table.

This billiard-theoretical approach gives an extremely simple way to solve the classic elastic collision problem, which was first posed by Sinai[103].

4.6.1 *Three particles on an infinite line*

Consider a three-particle system on an infinite line that consists of two approaching solid balls (also called cannonballs), each of mass M. Between them (non-symmetrically located) lies a ping-pong ball of mass m, where m is very small compared to M. When collisions occur between the cannonballs and the intervening ping-pong ball, the latter rattles back and forth with rapidly increasing speed until its momentum is sufficient to drive the cannonballs apart (see Fig. 4.17).

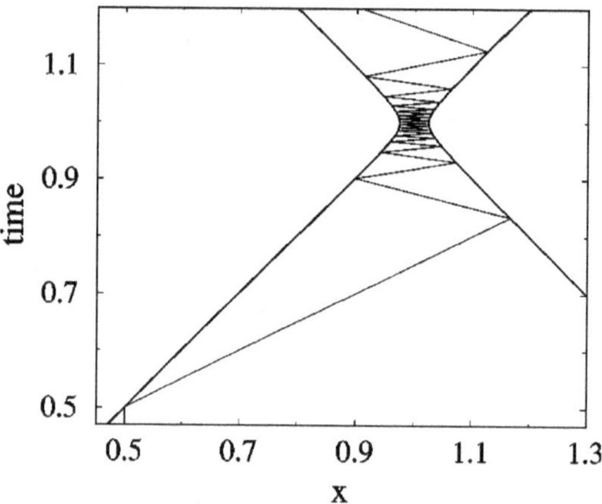

Fig. 4.17   Space-time diagram of the typical evolution of two cannonballs (bold lines) approaching an initially stationary ping-pong ball (light line). The cannonballs each have mass $M = 1$ and initial conditions $(x_1(0), v_1(0)) = (0, 1)$ and $(x_3(0), v_3(0)) = (2, -1)$. A ping-pong ball of mass $m = 0.005$ is initially at $x_2(0) = 1/2$. There are 31 collisions in total before the three particles recede. The first 30 collisions are shown.

In the final state, the three particles are receding from each other. How many collisions occur before this final state is reached? Using energy and momentum conservation, we can determine the state of the system after each collision and therefore find the number of collisions before the three particles mutually recede. However, this approach is complicated and provides minimal physical insight. Now we give a simpler solution by mapping the original three-particle system onto a billiard in an appropriately defined domain.

Let the coordinates of the balls (particles) be x_1, x_2, and x_3, with $x_1 < x_2 < x_3$. This order between the particles translates to a geometrical constraint on the accessible region for the billiard ball in the three-dimensional x_i space, i.e. the configuration space. Then the trajectories of the particles on the line translate to the trajectory $x_1(t), x_2(t), x_3(t)$ of a billiard ball in the allowed configuration space.

Introduce the rescaled coordinates $y_i = \sqrt{m_i} \cdot x_i$. These coordinates satisfy the conditions

$$\frac{y_1}{\sqrt{M}} < \frac{y_2}{\sqrt{m}}, \qquad \frac{y_2}{\sqrt{m}} < \frac{y_3}{\sqrt{M}}. \tag{4.5}$$

In y_i space, the constraints correspond, respectively, to the effective billiard ball being confined to the half-space to the right of the plane $y_1/\sqrt{M} = y_2/\sqrt{m}$ and to the half-space to the left of the plane $y_2/\sqrt{m} = y_3/\sqrt{M}$ see Fig. 4.18.

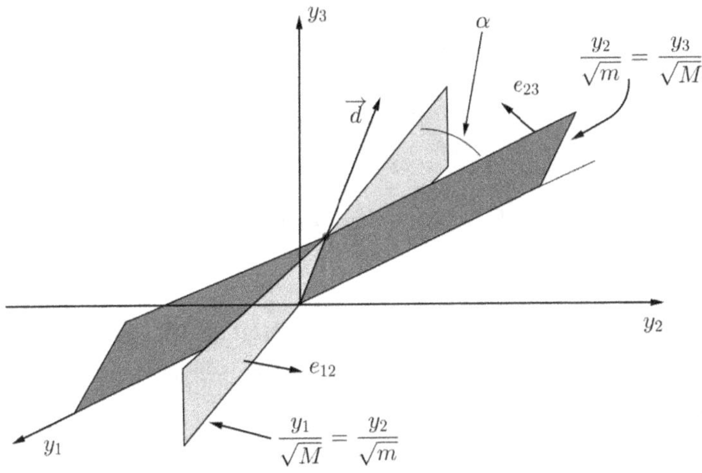

Fig. 4.18   The allowed wedge in the y_i coordinate system for a system of two cannonballs and an intervening ping-pong ball on an infinite line. The wedge is the region where the vectors e_{12} and e_{23} point toward.

These constraints define the allowed region as an infinite wedge of opening angle α. The use of rescaled coordinates ensures that all collisions between the effective billiard particle and these constraint planes are specular. Here the specular means that the angle of incidence equals the angle of reflection. Now momentum conservation gives

$$Mv_1 + mv_2 + Mv_3 = \sqrt{M}w_1 + \sqrt{m}w_2 + \sqrt{M}w_3 = 0, \qquad (4.6)$$

here, without loss of generality, we take the total momentum to be zero. In this zero momentum reference frame, the trajectory of the billiard ball is always perpendicular to the diagonal vector $\vec{\mathbf{d}} = (\sqrt{M}, \sqrt{m}, \sqrt{M})$. Consequently, we can reduce the three-dimensional billiard to a two-dimensional system in the plane perpendicular to $\vec{\mathbf{d}}$.

Now it remains to find the wedge angle α. The normals to the two constraint planes are

$$\mathbf{e}_{12} = \left(-\frac{1}{\sqrt{M}}, \frac{1}{\sqrt{m}}, 0\right), \quad \mathbf{e}_{23} = \left(0, -\frac{1}{\sqrt{m}}, \frac{1}{\sqrt{M}}\right).$$

Thus the angle between these planes is given by

$$\alpha = \arccos\left(-\frac{\mathbf{e}_{12} \cdot \mathbf{e}_{23}}{|\mathbf{e}_{12}|\,|\mathbf{e}_{23}|}\right) = \arccos\left(\frac{1}{1 + \frac{m}{M}}\right). \qquad (4.7)$$

By (4.7), in the limit $m/M \to 0$, we obtain $\alpha \approx \sqrt{\frac{2m}{M}}$. The maximum number $N_{\max}$ of possible collisions is determined by the number of wedges that fit into the half plane. This condition gives

$$N_{\max} = \frac{\pi}{\alpha} \approx \pi\sqrt{\frac{M}{2m}}. \qquad (4.8)$$

Therefore, if $m/M \to 0$, the opening angle of the wedge goes to zero and correspondingly, the number of collisions diverges.

4.6.2 *Triangular billiard: Three particles on a ring*

Following [28], [44], [89] consider three elastically colliding particles of arbitrary masses m_1, m_2, and m_3 on a finite ring of length L. Making an imaginary cut in the ring between particles 1 and 3, we can write the order constraints of the three particles as

$$x_1 < x_2, \ x_2 < x_3, \ x_3 < x_1 + L. \qquad (4.9)$$

Now we employ the rescaled coordinates $y_i = \sqrt{m_i}x_i$ to ensure that all collisions of the billiard ball with the domain boundaries in the y_i coordinates are specular. Then the first two constraints again confine the particle to be between the planes defined by the normal vectors

$$\mathbf{e}_{12} = \left(-\frac{1}{\sqrt{m_1}}, \frac{1}{\sqrt{m_2}}, 0\right), \quad \mathbf{e}_{23} = \left(0, -\frac{1}{\sqrt{m_2}}, \frac{1}{\sqrt{m_3}}\right).$$

By an offset of L we mean that one must translate this plane by a distance $L\sqrt{m_3}$ along y_3. Without the offset of L, the constraint $x_3 < x_1 + L$ corresponds to a plane that slices the $y_1 - y_3$ plane and passes through the origin. The fact that x_3 is the lesser coordinate also means that the billiard ball is confined to the near side of this constraint plane. Consequently the billiard ball must remain within a triangular bar whose outlines are shown in Fig. 4.19.

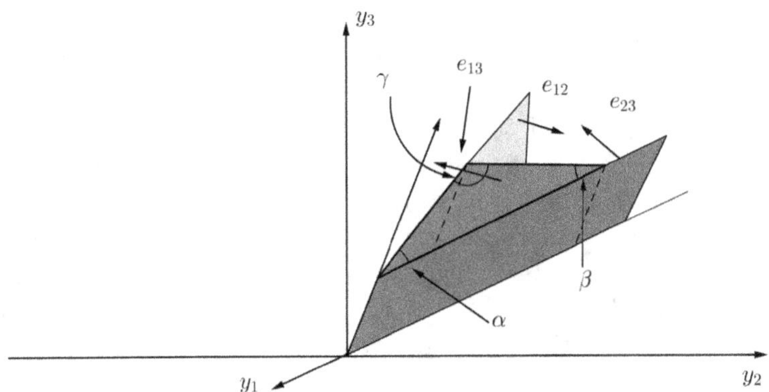

Fig. 4.19 The allowed region (configuration space) in the y_i coordinates for three particles of arbitrary masses on a ring of circumference L. The triangular billiard with angles α, β, and γ is defined by the thick solid lines.

Assuming that the total momentum of the system is zero, we get

$$(\sqrt{m_1}, \sqrt{m_2}, \sqrt{m_3}) \cdot (w_1, w_2, w_3) = 0,$$

and the trajectory of the billiard ball remains within a triangle perpendicular to the long axis of the bar, with angles α, β, and γ. Let us compute these angles by the same approach given in (4.7):

$$\alpha = \arccos\left(-\frac{\mathbf{e}_{12} \cdot \mathbf{e}_{23}}{|\mathbf{e}_{12}||\mathbf{e}_{23}|}\right) = \arccos\left(\sqrt{\frac{m_1 m_2}{(m_1 + m_2)(m_2 + m_3)}}\right). \quad (4.10)$$

The angles β and γ can be obtained by cyclic permutations of this formula.

Thus the elastic collisions of three particles on a finite ring can be mapped onto the motion of a billiard ball within a triangular billiard table. One can then exploit the wealth of knowledge about triangular billiards (Section 3.3 to infer basic collisional properties of the three-particle system). For example, periodic or chaotic behavior of the billiard translates to periodic or chaotic behavior in the three-particle collision sequence.

Let us give some applications of known results about billiards on triangular tables to the mechanical system of three elastic rods on a ring (see [44]):

i) Any acute triangular table admits orbits of period six. Three rods on a ring with any positive masses display analogous periodic motions. The minimal period is six, unless two balls collide when the third is at a specific position on the ring, e.g., if $x_2 = 0$ and $m_3 x_1 = m_1 x_3$. This special case corresponds to the pedal 3-orbit on an acute triangle, since any billiard orbit with odd period n is a limiting case of orbits with period $2n$ [40], [51].

ii) On an equilateral triangle there are any even periodic (exceptions are 2, 8, 12, and 20) billiard orbits [38]. Orbits with even periods correspond to periodic motions of three identical rings with arbitrary initial positions along the ring. If the angles of the billirad table are rational multiples of π, then the known billiard theorems (Chapter 3) apply to the rod problem. But rod masses corresponding to these rational triangles do not have apparent physical significance.

iii) The mechanical system of three elastic rods on a ring is typically ergodic [51], [69].

iv) All non-periodic orbits on any polygonal table come arbitrarily close to at least one vertex [40]. Consequently three rods on a ring in a non-periodic orbit must come arbitrarily close to a triple collision.

v) A generalization of the procedure maps the motion of $N + 1$ rods with any masses onto that of one ball in an elastically bounded N-dimensional simplex, thus offering an alternative picture of the multi-component Tonks gas [63].

Moreover rods moving on a ring can shed light on billiards. Denote by d the mean distance between ball-rail impacts along a billiard trajectory. For the equilateral triangle of side l, the equivalent rod problem makes it obvious that d depends on the initial direction of motion but not the initial position, and that

$$d = \frac{\sqrt{3}\,l}{4\cos\varphi},$$

where $\varphi \in [0, \frac{\pi}{6}]$ is the smallest of the angles between the velocity of the ball and the normals to the legs of the triangle.

Recall the Birkhoff's extremality property: every periodic trajectory on any convex table has an extremal (minimal or maximal) length among all closed polygonal lines inscribed in the table and with vertices close to the reflection points of the given trajectory. This fact implies that the extrema of d on any convex table correspond to periodic orbits. We obtain an orbit of period six for $d_{max} = \frac{1}{2}$ and period four for $d_{min} = \frac{\sqrt{3}l}{4}$. The geometric mean length of a randomly drawn chord of the triangle is $\frac{l\pi}{4\sqrt{3}}$ and lies between these extrema.

4.7 n-particle gas

Consider $n \geq 3$ balls with masses $m_1, m_2, \ldots, m_n$ which are moving on the x-axes. The state of the system is an vector $x = (x_1, x_2, \ldots, x_n) \in \mathbb{R}^n$ satisfying

$$x_1 \leq x_2 \leq \cdots \leq x_n.$$

Then for configuration space we have

$$Q = \{x \in \mathbb{R}^n : x_1 \leq x_2 \leq \cdots \leq x_n\}.$$

Similarly to the case $n = 2$, using $y_i = \sqrt{m_i} x_i$, $i = 1, \ldots, n$, one can reduce the system to a billiard system on the $n - 1$-hedral angle:

$$Q = \left\{ y \in \mathbb{R}^n : \frac{y_1}{\sqrt{m_1}} \leq \frac{y_2}{\sqrt{m_2}} \leq \cdots \leq \frac{y_n}{\sqrt{m_n}} \right\}.$$

This system can be considered as a system of gas particles on the real line. The question we are interested is

Question. For n particles (balls) on a straight line with masses m_i, $i = 1, \ldots, n$, with the initial positions $x_1, \ldots, x_n$ and initial speeds $v_1, \ldots, v_n$ to find out, whether always finite the number of collisions between them on an infinite period of time, counting all happening pairwise collisions (assuming that they satisfy conservation laws of impulse and energy).

The following proposition says that in case of equal masses, the answer to this question affirmative:

Proposition 4.6. *If $m_1 = m_2 = \cdots = m_n$ then the maximal number of collisions is $N_{max} = \frac{n(n-1)}{2}$.*

Proof. Consider Otx system of coordinate, then the trajectory of each ball, say i, can be considered as a ray (half line) started on $x(0) = a_i$ and angular coefficient v_i: $x_i(t) = v_i t + a_i$. The set of graphs of movements of n balls of equal masses represents union of n rays on the Otx plane (see Fig. 4.20).

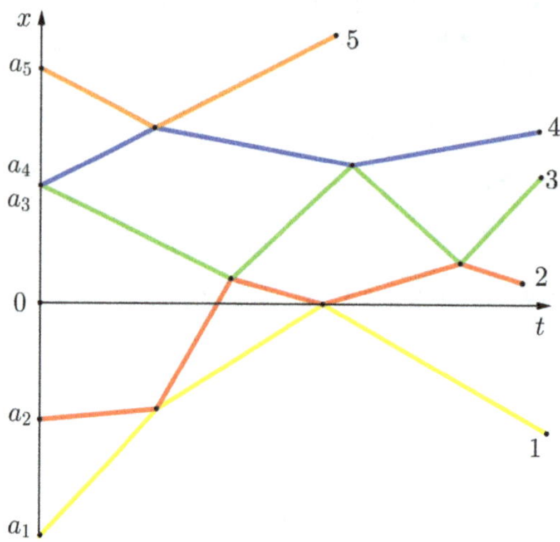

Fig. 4.20   The configuration of movement of n-balls with equal mass on the x-axes.

By this figure it is easy to track the schedule of $x = x_i(t)$. From the figure it is visible that the first ball after scattering has speed v_3, the second - v_5, third - v_2, the fourth - v_1 and the fifth - v_4. It is also visible that the maximum number $N_{\max}$ of impacts of n balls of equal masses is equal to maximum number of points of pairwise crossings of n rays on the plane. As each two rays can be crossed only in one point, we have

$$N_{\max} = \binom{n}{2} = \frac{n(n-1)}{2}.$$

Physically the number $N_{\max}$ can be obtained, for example, if the balls move from left to right, and the initial speed of each ball is more than the speed of his right neighbor. Let's note, that during collision of two identical balls they exchange with speeds, or, speaking differently, one ball passes through another, having exchanged their numbers (on it each ball can "pass through another" only no more than $n - 1$ times).   □

Now we consider the case $n \geq 3$ balls with arbitrary masses.

Theorem 4.4. *For n balls on a straight line with arbitrary masses m_i, $i = 1, \ldots, n$, the number of collisions between them is finite on an infinite period of time.*

Proof. We use induction over $n \geq 3$. For the case $n = 3$ we have the number of pairwise collisions is equal to (see (4.10))

$$N(m_1, m_2, m_3) = -\left\lfloor -\frac{\pi}{\alpha} \right\rfloor$$

$$= -\left\lfloor -\frac{\pi}{\arccos \sqrt{m_1 m_3 / [(m_1 + m_2)(m_2 + m_3)]}} \right\rfloor.$$

Note that this number is sharp (there is a physical configuration of a movement of the gas for which this number obtained). Moreover, in case of equal masses, we have $N(m, m, m) = 3$.

Assume now (as for the mathematical induction method) that for any initial k balls with $k < n$, during the time interval from 0 until T (where T may be $+\infty$) there are only a finite number of collisions. We shall show that the number of collisions is finite at $k = n$ too. Suppose that is not true, i.e., there is some initial configuration of n balls which have collisions infinitely many times during the time from 0 until T. We shall show that this leads to a contradiction. We come to the contradiction by making the following four steps (assuming that there are infinitely many collisions):

Step 1. *Existence of limits.* We numerate the balls in the order from the right to the left as shown in Fig. 4.21.

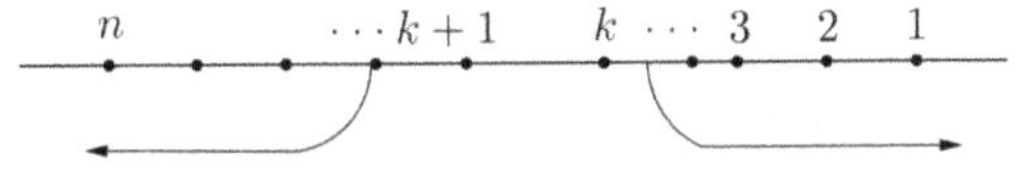

Fig. 4.21 The increasing impulse p_k.

In each time moment we consider sum of impulses of k right balls

$$p_k = m_1 v_1 + m_2 v_2 + \cdots + m_k v_k,$$

where $k = 1, \ldots, n$. We show that at each pairwise collision the impulse p_k may only increase. Indeed, the impulse p_k does not change if the balls staying on the left from $k + 1$-th ball or ones staying on the right from the k-th ball will interact (collisions). This is because the balls $k+2, k+3, \ldots, n$ do not present in sum p_k and the speed v_k and the impulse $p_{k-1} = m_1 v_1 +

$\cdots + m_{k-1}v_{k-1}$ are unchanged (see Fig. 4.21). If the k-th and $k+1$-th balls interact then p_k increases: the $k+1$-th ball will shift the k-th bal to the right and will increase the speed v_k, thus it increases the sum p_k.

Now let us numerate all collisions of the system of balls by natural numbers: $1, 2, \ldots, s, \ldots$. Denote by

$v_i^{(s)}$ the speed of the i-th ball after s-th (by the above order) collision;

$p_k^{(s)}$ the impulse p_k after s-th (by the above order) collision.

By the above mentioned result we have for any $k = 1, 2, 3, \ldots$ that

$$p_k^{(1)} \le p_k^{(2)} \le \cdots \le p_k^{(s)} \le \cdots \tag{4.11}$$

Moreover, by the conservation law of the kinetic energy of n balls we have

$$\frac{1}{2}\left(m_1 v_1^2 + m_2 v_2^2 + \cdots + m_n v_n^2\right) = \text{const} = E.$$

Therefore the speed of each ball i is bounded:

$$m_i v_i \le 2E, \quad |v_i| \le \sqrt{\frac{2E}{m_i}} \le \sqrt{\frac{2E}{m_{\min}}},$$

where

$$m_{\min} = \min\{m_1, m_2, \ldots, m_n\}.$$

Now for any k and s we have

$$p_k^{(s)} \le \sum_{i=1}^{k} m_i |v_i^{(s)}| \le \sqrt{\frac{2E}{m_{\min}}} \sum_{i=1}^{k} m_i. \tag{4.12}$$

By (4.11) and (4.12) we see that the sequences $\{p_k^{(s)}\}_{s=1}^{\infty}$, $k = 1, 2, \ldots$ are monotone and bounded. Consequently, each of such sequence has a limit, that is

$$\lim_{s \to \infty} p_k^{(s)} = \tilde{p}_k, \quad k = 1, 2, \ldots, n.$$

Now we show that the following limit exists

$$\lim_{s \to \infty} v_k^{(s)} = w_k, \quad k = 1, 2, \ldots, n. \tag{4.13}$$

To prove the existence of the limit (4.13) we use the mathematical induction over k. For the $k = 1$ the existence follows from the equality $p_1^{(s)} = m_1 v_1^{(s)}$, because $p_1^{(s)}$ has a limit. Now since the existence of limits of $v_1^{(s)}$ and $p_2^{(s)}$ known, the sequence $v_2^{(n)}$ has limit because

$$p_2^{(s)} = m_1 v_1^{(s)} + m_2 v_2^{(s)}.$$

Assume now that $v_i^{(s)}$ has limit for each $i = 1, \ldots, n-1$, then existence of the limit of $v_n^{(s)}$ follows from the following

$$m_n v_n^{(s)} = p_k^{(s)} - \sum_{i=1}^{n-1} m_i v_i^{(s)}.$$

Step 2. *Equality of the limiting speeds.* In this step we prove that (see (4.13))

$$w_1 = w_2 = \cdots = w_n = w. \tag{4.14}$$

Consider two balls numbered by k and $k+1$. By our assumption (in the beginning part of the proof) there are infinitely many collisions of n balls during time from 0 until T, then the balls k and $k+1$ also have infinitely many collisions. Indeed, if they had only finitely many collisions then there is a time t_0, $0 < t_0 < T$ such that the k-th and $k+1$-th balls will not interact at t_0 and after that time. Then, at time t_0, the system of n balls will be separated to two subsystems: right (consisting balls with numbers $1, 2, \ldots, k$) and left (consisting balls with numbers $k+1, k+2, \ldots, n$). Each of the subsystem consists strictly less than n balls. Therefore, by the assumption of the induction, we have a finite number of collisions in each subsystem, as during time 0 to t_0, as well as during the time t_0 to T. Consequently, we have that the initial system of n balls has also finitely many collisions, which contradicts our assumption.

Now consider some collisions of the balls k and $k+1$ at s, then by the conservation law of impulse we have

$$m_k v_k^{(s)} + m_{k+1} v_{k+1}^{(s)} = m_k v_k^{(s-1)} + m_{k+1} v_{k+1}^{(s-1)}.$$

Moreover, the conservation law of energy gives

$$m_k (v_k^{(s)})^2 + m_{k+1} (v_{k+1}^{(s)})^2 = m_k (v_k^{(s-1)})^2 + m_{k+1} (v_{k+1}^{(s-1)})^2.$$

From the obtained equalities we get

$$m_k (v_k^{(s)} - v_k^{(s-1)}) = m_{k+1} (v_{k+1}^{(s-1)} - v_{k+1}^{(s)}). \tag{4.15}$$

$$m_k ((v_k^{(s)})^2 - (v_k^{(s-1)})^2) = m_{k+1} ((v_{k+1}^{(s-1)})^2 - (v_{k+1}^{(s)})^2). \tag{4.16}$$

Since we are considering a collision of k-th and $k+1$-th balls at s-th interaction, we have

$$v_k^{(s)} \neq v_k^{(s-1)}, \quad v_{k+1}^{(s)} \neq v_{k+1}^{(s-1)}.$$

Therefore we can divide both sides of (4.16) correspondingly to (4.15) and get

$$v_k^{(s)} + v_k^{(s-1)} = v_{k+1}^{(s-1)} + v_{k+1}^{(s)}. \tag{4.17}$$

Since the balls k and $k+1$ have infinitely many collisions, we can take the limit from both sides of (4.17) when $s \to \infty$. Moreover, by existence of the limits we get $2w_k = 2w_{k+1}$, i.e., $w_k = w_{k+1}$. Hence (4.14) holds.

Step 3. *The computation of w.* Consider initial speeds of balls: $v_1^{(0)}, \ldots, v_n^{(0)}$. By the conservation law of impulse we have

$$m_1 v_1^{(0)} + \cdots + m_n v_n^{(0)} = m_1 v_1^{(s)} + \cdots + m_n v_n^{(s)}.$$

Then for the limit point w we have

$$m_1 v_1^{(0)} + \cdots + m_n v_n^{(0)} = (m_1 + \cdots + m_n)w,$$

i.e.

$$w = \frac{m_1 v_1^{(0)} + \cdots + m_n v_n^{(0)}}{m_1 + \cdots + m_n}.$$

The right side is the speed of the center of mass of the system of n balls. So, the speed of each ball converges to the speed of the center of mass of the system (if the number of collisions is infinite). Let us pass in this regard into the system of the center of masses, i.e. into a system of coordinates which moves uniformly with respect to initial system with speed w. In the system of the center of masses all limiting speeds are equal to 0. Therefore, the limiting energy is 0 in the system of the center of masses.

Step 4. *Where is the contradiction?* Now owing to inertness of the system connected with the center of masses, as well as in initial system, the law of energy conservation is executed. Since the limiting energy is equal 0, the initial energy of balls is also equal to 0 in the system of the center of masses. This can be only when all initial speeds of balls were equal 0 (in the system of the center of masses), i.e. when all initial speeds of balls were equal to w in an initial reference system. But at identical initial speeds of balls any collisions between them can not occur. It contradicts the initial assumption that in the system would occur infinitely many collisions. Therefore, the initial assumption is incorrect - the number of collisions can be only finitely many. This completes the proof. $\square$

On the real line consider n balls with arbitrary masses $m_1, m_2, \ldots, m_n$. Denote by $N(m_1, \ldots, m_n)$ the number of collisions between these balls. In [37] using theory of billiards in the $n-1$-hedral angle $\mathcal{Q}$ the following theorem is proved

Theorem 4.5. *([37], p. 178) The number N satisfies the following inequality*

$$N(m_1, \ldots, m_n) \le 2 \left(8n^2(n-1) \frac{m_{\max}}{m_{\min}} \right)^{n-2},$$

where

$$m_{\min} = \min\{m_1, \ldots, m_n\}, \quad m_{\max} = \max\{m_1, \ldots, m_n\}.$$

4.8   Broken ray tomography

In this section we discuss a surprising application of mathematical billiards.

Following [60], [61] we consider a ray of light traveling to the right on the real axis. Denote by $I(x)$ the intensity of the light at a point x and assume the material on the real axis has a non-constant attenuation coefficient $f \ge 0$. Then the following differential equation should be satisfied

$$I'(x) = -f(x)I(x). \tag{4.18}$$

For a given initial intensity $I(0)$ and the intensity at a distance x, one may easily find that

$$\log \left(\frac{I(0)}{I(x)} \right) = \int_0^x f(t)dt. \tag{4.19}$$

Thus by measuring the ratio of initial and final intensity of a beam of light we in effect measure the integral of the attenuation coefficient over its trajectory. This is true even when the trajectory is more complicated than a segment of the real line.

The fundamental problem of (scalar) ray tomography is to deduce knowledge of the attenuation coefficient f from its integrals over a suitable set of trajectories. Different materials and wavelengths have different attenuation coefficients, so any information about f can be translated into understanding of the structure of the object being imaged.

Consider a bounded C^1 domain Ω, (with boundary $\partial\Omega$ and $\overline{\Omega} = \Omega \cup \partial\Omega$) in $\mathbb{R}^n$ and a continuous attenuation coefficient $f : \overline{\Omega} \to \mathbb{R}$.

Definition 4.1. [60] A piecewise linear path in $\overline{\Omega}$ with reflections at the boundary as in geometric optics (the angle of incidence equals the angle of reflection, i.e. billiard's law) is called a broken ray.

Assume now that a device E is given, which is a nonempty subset of the boundary $\partial\Omega$. Suppose from E we emit light and measure the intensity of incoming light. Then the rest of the boundary, $\partial\Omega \setminus E$, acts as a reflector and consider those broken rays which start and end at E, which is called the set of tomography. The corresponding set of integrals of f over such broken rays is called the broken ray transform of f with the set of tomography E. We assume that all broken rays start and end at the set of tomography.

The main question is:

Question. If the integral of f is known over all broken rays with both endpoints in E, can the function f be fully reconstructed? How does the result depend on the domain Ω, the set of tomography E, and regularity assumptions on the unknown function f?

We note that if $E = \partial\Omega$ then broken rays are intersections of lines with the domain Ω. Therefore, by extending the unknown function f by zero to $R^n \setminus \Omega$, the problem reduces to the classical X-ray transform. For this case the answer to the uniqueness question is affirmative, and also regularity and feasible reconstruction algorithms are known (see [56], [84]).

In this section we only consider one example of the tomography. For more theory of this subject see [60], [61], [62] and the references therein.

Consider the case where Ω is the unit disk $D \subset \mathbb{R}^2$. Use polar coordinates (r, v) in the unit disk D and identify points on the boundary ∂D with the corresponding angle. Take $E = \{0\}$ as one point at angle zero.

A broken ray γ is characterized by the number of line segments n_γ, where there are $n_\gamma - 1$ reflections, the initial point $i_\gamma \in E$, the final point $\kappa_\gamma \in E$, the angle α_γ at which a line segment appears when seen from the origin, and an integer winding number m_γ such that the following the trajectory condition is satisfied

$$\nu_\gamma \alpha_\gamma = \kappa_\gamma - i_\gamma + 2\pi m_\gamma.$$

Thus these five parameters $n_\gamma, m_\gamma, \alpha_\gamma, i_\gamma, \kappa_\gamma$ uniquely define the broken ray γ. Here we need to the condition that $n_\gamma > 0$, but the other parameters may be negative. Denote by $z_\gamma = \cos(\alpha_\gamma/2)$ and $d = 2|\sin(\alpha_\gamma/2)|$ the distance from the origin to the trajectory and the length of each individual line segment. The parameters describing the broken ray and the billiard trajectory are illustrated in Fig. 4.22.

Theorem 4.6. [60] *If $f : \overline{D} \to \mathbb{R}$ is continuous and the set of tomography is a singleton, then integrals over broken rays uniquely determine the integral of f over any circle centered at the origin.*

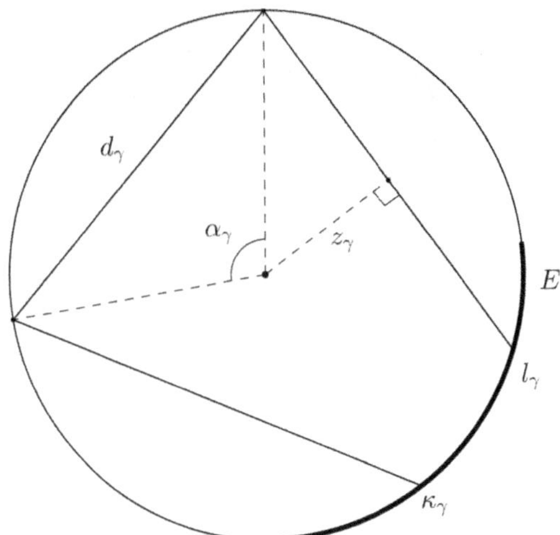

Fig. 4.22   An example of a broken ray γ in the disk. Here $m_\gamma = 1$ and $n_\gamma = 3$. The parameters $\alpha_\gamma, z_\gamma, d_\gamma$ and κ_γ as well as the set of tomography E are illustrated in the figure. The angles l_γ and κ_γ are identified with the corresponding points on the boundary.

Corollary 4.1. [60] *Let B^n denote the unit ball in $\mathbb{R}^n$. If $n \geq 2$ and $f : \overline{B}^n \to \mathbb{R}$ is continuous, and the set of tomography is a singleton, then integrals over broken rays uniquely determine the integral of f over any circle centered at the origin with the singleton in the circles plane.*

Bibliographical notes. In this chapter I tried to give simple applications of the billiard theory to physics. This chapter is based on [2], [11], [14], [28], [23], [33], [37], [38], [40], [41], [51], [60],[61], [62], [69], [87], [89], [103], [110], [111], [112] and many internet sources.

Bibliography

[1] P. Albers, S. Tabachnikov, *Introducing symplectic billiards.* Adv. Math. **333** (2018), 822-867.

[2] N.N. Andreev, S.P. Konovalov, N.M. Panyunin (Eds), *Mathematical component*, Mathematical Etudes. Moscow, 2015.(in Russian)

[3] T. Andreescu, O. Mushkarov, L. Stoyanov, *Geometric problems on maxima and minima.* Birkhäuser Boston, Inc., Boston, MA, 2006.

[4] E.G. Altmann, J.C. Leito, J.V. Lopes, *Effect of noise in open chaotic billiards.* Chaos **22**(2) (2012), 026114, 7 pp.

[5] V. Arnold. *Mathematical methods of classical mechanics*, Springer-Verlag, 1989.

[6] M. Arnold, M. Bialy, *Nonsmooth convex caustics for Birkhoff billiards.* Pacific J. Math. **295**(2) (2018), 257-269.

[7] A. Balitskiy, *Shortest closed billiard trajectories in the plane and equality cases in Mahler's conjecture.* Geom. Dedicata **184** (2016) 121-134.

[8] B. Batistic, M. Robnik, *Fermi acceleration in time-dependent billiards: theory of the velocity diffusion in conformally breathing fully chaotic billiards.* J. Phys. A **44**(36) (2011), 365101, 21 pp.

[9] P. Beckmann. *A history of* π. Golem Press. 1971.

[10] V. Benci, F. Giannoni. *Periodic bounce trajectories with a low number of bounce points.* Ann. Inst. Poincare, Anal. Non Lineaire 6 (1989), 73-93.

[11] M. Berry, M. Robnik. *Classical billiards in magnetic fields.* J. Phys. A 18 (1985), 1361-1378.

[12] C.D. Birkhoff, *Proof of the ergodic theorem.* Proc. Natl. Acad. Sci. USA. **17**(12) (1931), 656-660.

[13] J. Bobok, S. Troubetzkoy, *Code and order in polygonal billiards.* Topology Appl. **159**(1) (2012), 236-247.

[14] M. Boshernitzan, G. Galperin, T. Krüger, S.Troubetzkoy. *Some remarks on periodic billiard orbits in rational polygons.* Preprint. 1994. https://www.math.stonybrook.edu/preprints/ims94-14.pdf

[15] M. Boshernitzan, G. Galperin, T. Krüger, S.Troubetzkoy. *Periodic billiard orbits are dense in rational polygons.* Trans. Amer. Math. Soc. **350**(9) (1998), 3523-3535.

[16] I. Bouw, M. Möller, *Teichmüller curves, triangle groups, and Lyapunov exponents*, Annals of Mathematics, **172**(1) (2010), 139-185.

[17] M. Braun. *Differential equations and their applications: An introduction to applied mathematics.* Springer, 4th edition, 1992.

[18] L.A. Bunimovich, *Mechanisms of chaos in billiards: dispersing, defocusing and nothing else.* Nonlinearity. **31**(2) (2018), R78-R92.

[19] L.A. Bunimovich, A. Grigo, *Focusing components in typical chaotic billiards should be absolutely focusing.* Comm. Math. Phys. **293**(1) (2010), 127-143.

[20] Yu. Burago, V. Zalgaller. *Geometric inequalities*, Springer-Verlag, 1988.

[21] J.M. Casas, M. Ladra, U.A. Rozikov, *A chain of evolution algebras*, Linear Algebra Appl. **435**(4) (2011) 852–870.

[22] D. Castro, *Corrections.* Quantum Magazine. Washington DC: Springer-Verlag. **7**(3) (1997) 42-48.

[23] N. Chernov, G. Galperin, *Search light in billiard tables.* Regul. Chaotic Dyn. **8**(2) (2003), 225-241.

[24] N. Chernov, R. Markarian. *Introduction to the Ergodic Theory of Chaotic Billiards.* Monographs of the Institute of Math. and Rel. Sci.], 19. IMCA, Lima; Pontificia Univ. Catlica del Perú, Lima, 2001.

[25] N. Chernov, R. Markarian, *Chaotic billiards.* Mathematical Surveys and Monographs, 127. American Mathematical Society, Providence, RI, 2006.

[26] B. Cipra, R. Hanson, A. Kolan. *Periodic trajectories in right triangle billiards.* Phys. Rev. E 52 (1995), 2066-2071.

[27] I. P. Cornfeld, S. V. Fomin, Ya. G. Sinai, *Ergodic theory*, Springer- Verlag, 1982.

[28] S. G. Cox, G. J. Ackland, *How efficiently do three pointlike particles sample phase space?*, Phys. Rev. Lett. **84** (2000), 2362-2365.

[29] D. Davis, *Billiards and flat surfaces*, Snapshots of modern mathematics from Oberwolfach. 1 (2015), 1–9.

[30] D. Davis, *Geometry, Surfaces and Billiards.* Williamstown, MA, 2016. `https://math.williams.edu/files/2016/05/424-problems-all.pdf`

[31] A. Deniz, A.V. Ratiu, *On the existence of Fagnano trajectories in convex polygonal billiards.* Regul. Chaotic Dyn. **14**(2) (2009), 312-322.

[32] A. Deniz, A.V. Ratiu, *Periodic trajectories in some cyclic polygons.* J. Geom., **86** (2006), 185-186.

[33] R. L. Devaney, *An introduction to chaotic dynamical system*, Westview Press, (2003).

[34] A. Dhar, R. Madhusudhana, N. Shankar, S. Sridhar, *Isospectrality in chaotic billiards.* Phys. Rev. E (3) **68**(2) (2003), 026208, 5 pp.

[35] S. N. Elaydi, *Discrete chaos*, Chapman Hall/CRC, (2000).

[36] O. Galor, *Discrete dynamical systems.* Springer, Berlin, 2007.

[37] G. Galperin, A. Zemlyakov, *Mathematical billiards.* Nauka, Moscow, 1990 (in Russian).

[38] G. A. Galperin, N. I. Chernov, *Billiards and chaos.* Znanie, Moscow, 1991 (in Russian).

[39] G. Galperin, A. Stepin, Ya. Vorobets. *Periodic billiard trajectories in poly-*

gons: generating mechanisms. Russ. Math. Surv. 47(3) (1992), 5-80.

[40] G. Galperin, T. Krüger, S. Troubetskoy, *Local instability of orbits in polygonal and polyhedral billiards*, Commun. Math. Phys. **169** (1995), 463.

[41] G. Galperin, D. Zvonkine. *Periodic billiard trajectories in right triangles and right-angled tetrahedra.* Regul. Chaotic Dyn. **8**(1) (2003), 29-44.

[42] G. Galperin, *Playing pool with π (the number π from a billiard point of view).* Regul. Chaotic Dyn. **8**(4) (2003), 375-394.

[43] R.N. Ganikhodzhaev, F. M. Mukhamedov, U.A. Rozikov, *Quadratic stochastic operators and processes: results and open problems.* Inf. Dim. Anal. Quant. Prob. Rel. Fields. **14**(2) (2011), 279-335.

[44] S. L. Glashow, L. Mittag, *Three rods on a ring and the triangular billiard,* J. Stat. Phys. **87** (1996), 937-941.

[45] G. Harris, *Polygonal Billiards.* `http://www.siue.edu/~aweyhau/teaching/seniorprojects/harris_final_042907.pdf`

[46] G. Huang, V. Kaloshin, A. Sorrentino, *On the marked length spectrum of generic strictly convex billiard tables.* Duke Math. J. **167**(1) (2018), 175-209.

[47] T. Gorin, *Generic spectral properties of right triangle billiards.* J. Phys. A **34**(40) (2001), 8281-8295.

[48] E. Gutkin. *Billiard dynamics: a survey with the emphasis on open problems.* Regul. Chaotic Dyn. **8** (2003), 1-13.

[49] E. Gutkin. *Billiard dynamics: an updated survey with the emphasis on open problems.* `https://arxiv.org/pdf/1301.2547.pdf`

[50] E. Gutkin. *Problems on billirads,* Preprint no 103. 2001. Max Planck Institute. Leipzig.

[51] E. Gutkin, *Billiards in polygons: Survey of recent results,* J. Stat. Phys. **81**(7) (1996),

[52] E. Gutkin, S. Troubetzkoy, *Directional flows and strong recurrence for polygonal billiards,* Pittman Res. Notes in Math. **362** (1997), 21-45.

[53] M. Freiberger. *Chaos on the billiard table,* `https://plus.maths.org/content/chaos-billiard-table`

[54] J. Hale. *Ordinary Differential Equations.* Krieger Publishing Company, 2nd edition, 1980.

[55] G. H. Hardy and E. M. Wright, *An Introduction to the Theory of Numbers.* Oxford University Press, Oxford, 1979.

[56] S. Helgason, *The Radon Transform,* 2. edn., Birkhüser, 1999.

[57] F. Holt. *Periodic reflecting paths in right triangles.* Geom. Dedicata **46** (1993), 73-90.

[58] W. P. Hooper, *Periodic billiard paths in right triangles are unstable.* Geom. Dedicata **125**(1) (2007), 39-46.

[59] W. P. Hooper, R. E. Schwartz, *Billiards in nearly isosceles triangles.* Journal of Modern Dynamics, **3**(2) (2009), 159-231.

[60] J. Ilmavirta, *Broken ray tomography in the disc.* Inverse Problems **29**(3) (2013), 035008, 17 pp.

[61] J. Ilmavirta, *On the broken ray transform.* Thesis. University Printing House. Jyväskylä. 2014.

[62] J. Ilmavirta, S. Mikko, *Broken ray transform on a Riemann surface with a convex obstacle.* Comm. Anal. Geom. **24**(2) (2016), 379-408.

[63] D. W. Jepson, *Dynamics of a simple manybody system of hard rods,* J. Math. Phys. **6** (1965), 405.

[64] V. Kaloshin, A. Sorrentino, *On the local Birkhoff conjecture for convex billiards.* Ann. of Math. (2) **188**(1) (2018), 315-380.

[65] A. B. Katok, B. Hasselblatt, *Introduction to the modern theory of dynamical systems,* Cambridge University Press, 1995.

[66] A.B. Katok, *Billiard table as a mathematicians playground,* Surveys in Modern Mathematics, Cambridge Univ. Press, 2005, pp. 216-242.

[67] S.H. Kellert, *In the wake of chaos: Unpredictable order in dynamical systems.* University of Chicago Press. 1993, p. 32.

[68] R. Kenyon, J. Smillie, *Billiards on rational-angled triangles,* Comment. Math. Helv. **75** (2000), 65-108.

[69] S. Kerckhoff, H. Masur, J. Smillie, *Ergodicity of billiard flows and quadratic differentials,* Aml. Math. **124** (1986), 293.

[70] V.B. Kokshenev, *On chaotic dynamics in rational polygonal billiards.* SIGMA Symmetry Integrability Geom. Methods Appl. 1 (2005), Paper 014, 9 pp.

[71] H.J. Korsch, F. Zimmer, *Chaotic billiards.* Computational statistical physics, 15-36, Springer, Berlin, 2002.

[72] V.V. Kozlov, D.V. Treshchev, *Billiards: A Genetic Introduction to the Dynamics of Systems with Impacts,* American Mathematical Society, Translations of Mathematical Monographs, vol. 89., 1991.

[73] M. Ladra, U.A. Rozikov, *Construction of flows of finite-dimensional algebras.* Jour. Algebra. **492** (2017), 475-489.

[74] V. Lazutkin. *The existence of caustics for a billiard problem in a convex domain.* Math. USSR, Izvestija **7** (1973), 185-214.

[75] N.J. Lennes, *On the motion of a ball on a billiard table.* Amer. Math. Monthly, **12**(8-9) (1905), 152-155.

[76] S.A. Málaga, S. Troubetzkoy, *Weakly mixing polygonal billiards.* Bull. Lond. Math. Soc. **49**(1) (2017), 141-147.

[77] H. Masur, *Closed trajectories for quadratic differentials with an application to billiards,* Duke Math. J., **53**(2) (1986), 307-314.

[78] H. Masur, S. Tabachnikov. *Rational billiards and flat structures.* Handbook in Dynamical Systems Vol. 1A, Elsevier, 2002, p. 1015-1089.

[79] J. Mather. *Non-existence of invariant circles.* Ergod. Th. Dyn. Syst. **4** (1984), 301-309.

[80] C.D. Mayer. *Matrix analysis and applied linear algebra.* SIAM, 2000.

[81] I. Mikoss, P. Garcha, *An exact map for a chaotic billiard.* Internat. J. Modern Phys. B **25**(5) (2011), 673-681.

[82] J. Moser, *On invariant curves of area-preserving mappings of an annulus.* Nachr. Akad. Wiss. Göttingen Math.-Phys. Kl. II **1962** (1962), 1-20.

[83] J. Moser, E. J. Zehnder. *Notes on Dynamical Systems.* Courant Lecture Notes in Mathematics. AMS. 2005.

[84] F. Natterer, *The Mathematics of Computerized Tomography,* Society for

Industrial and Applied Mathematics, 2001.

[85] Z. Nitecki. *Differentiable dynamics,* MIT Press, Cambridge, MA. 1971.

[86] S. W. Park. *An Introduction to dynamical billiards.* `https://math.uchicago.edu/~may/REU2014/REUPapers/Park.pdf`

[87] R. Peirone. *Reflections can be trapped.* Amer. Math. Monthly, **101** (1994), 259-260.

[88] D. Pinheiro. *Notes on continuous-time dynamical systems.* Preprint. 2011.

[89] S. Redner. *A billiard-theoretic approach to elementary one-dimensional elastic collisions.* Amer. J. Phys. **72** (2004), 1492-1498.

[90] U.A. Rozikov. *Gibbs measures on Cayley trees.* World Sci Publ, Singapore. 2013.

[91] U.A. Rozikov. *Mathematical billiards.* Asia Pacific Math. Newsl. **2**(2) (2012), 19-23.

[92] U.A. Rozikov. *What is mathematical billiard?* Math. Track. **3** (2007), 56-65.

[93] U.A. Rozikov, U.U. Zhamilov, *On dynamics of strictly non-Volterra quadratic operators on two-dimensional simplex.* Sbornik: Math. **200**(9) (2009), 1339-1351.

[94] U.A. Rozikov, U.U. Zhamilov, *On F-quadratic stochastic operators.* Math. Notes. **83**(4) (2008), 554-559.

[95] R.C. Robinson. *An introduction to dynamical systems, continuous and discrete.* Pearson Edu. Inc. 2004.

[96] R. Samajdar, S.R. Jain, *Nodal domains of the equilateral triangle billiard.* J. Phys. A **47**(19) (2014), 195101, 22 pp.

[97] L. Santalo. *Integral geometry and geometric probability,* Addison- Wesley, 1976.

[98] R. E. Schwartz. *Obtuse triangular billiards I: Near the (2,3,6) triangle.* Journal of Experimental Mathematics, **15**(2) (2006), 161-182.

[99] R. E. Schwartz. *Obtuse triangular billiards II: 100 degrees worth of periodic trajectories.* Journal of Experimental Mathematics, **18**(2) (2008), 137-171.

[100] A.M. Sharkovskii, *Co-existence of cycles of a continuous mapping of the line into itself.* Ukrainian Math. J. **16** (1964), 61-71.

[101] H. S. Shultz, R.C. Shiflett, *Mathematical Billiards.* The Mathematical Gazette, **72**(460) (1988), 95-97.

[102] Ya. G. Sinai, *Dynamical Systems with elastic reflections,* Russian Math. Surv., **25**, (1970) 137-191.

[103] Ya. G. Sinai. *Introduction to ergodic theory.* Princeton University Press, Princeton, N.J. 1976.

[104] Ya. G. Sinai. *Hyperbolic billiards,* Proc. ICM, Kyoto 1990, Math. Soc. Japan, Tokyo, 1991, pp. 249-260.

[105] A. Skripchenko, S. Troubetzkoy, *Polygonal billiards with one sided scattering.* Ann. Inst. Fourier (Grenoble) **65**(5) (2015), 1881-1896.

[106] J. Smillie, C. Ulcigrai, *Beyond Sturmian sequences: coding linear trajectories in the regular octagon.* Proc. Lond. Math. Soc. (3), **102**(2) (2011), 291-340.

[107] D. Szász (Editor), *Hard ball systems and the Lorentz gas*, Springer Verlag, Berlin, 2000. (Encyclopedia of Mathematical Sciences, 101, Mathematical Physics, II.)

[108] S. Tabachnikov. *Billiards*, Societe Mathematique de France. "Panoramas et Syntheses," No. 1., 1995.

[109] S. Tabachnikov, *Fagnano orbits of polygonal dual billiards*, Geom. Dedicata, **77** (1999), 279-286.

[110] S. Tabachnikov, *Geometry and billiards*. Student Mathematical Library, 30. American Mathematical Society, Providence, RI; Mathematics Advanced Study Semesters, University Park, PA, 2005.

[111] S. Tabachnikov, *Remarks on magnetic flows and magnetic billiards, Finsler metrics and a magnetic analog of Hilberts fourth problem*, Dynamical systems and related topics, Cambridge Univ. Press, 2004, pp. 233-252.

[112] G. Tokarsky. *Polygonal rooms not illuminable from every point*. Amer. Math. Monthly, **102** (1995), 867-879.

[113] G. Tokarsky, *Galperin's triangle example*. Comm. Math. Phys. **335**(3) (2015), 1211-1213.

[114] S. Troubetzkoy, *Periodic billiard orbits in right triangles*. Ann. Inst. Fourier (Grenoble) **55**(1) (2005), 29-46.

[115] S. Troubetzkoy, *Dual billiards, Fagnano orbits, and regular polygons*. Amer. Math. Monthly, **116**(3) (2009), 251-260.

[116] C. Ulcigrai, *Mathematical billiards, flat surfaces and dynamics*. Prospects in Mathematics, Edinburgh, 17-18 December 2010.

[117] W. Veech, *Teichmüller curves in moduli space, Eisenstein series and an application to triangular billiards*, Inventiones Mathematicae, **87** (1989), 553-583.

[118] Ya. B. Vorobets, G.A. Galperin, A. M. Stëpin,*Periodic billiard trajectories in polygons: generation mechanisms*, Russian Math. Surveys **47**(3) (1992) 5-80.

[119] C. Ward, *Calculation of Fuchsian groups associated to billiards in a rational triangle*, Ergodic Theory and Dynamical Systems, **18**(4) (1998), 1019-1042.

[120] M. Wojtkowski, *Two applications of Jacobi fields to the billiard ball problem*. J. Diff. Geom. **40** (1994), 155-164.

[121] Y. Yang, *Stable billiard paths on polygons*, Undergraduate Honor Thesis, Brown University. 2013.

[122] Y. Yang, *Upper triangular matrices and billiard arrays*. Linear Algebra Appl. **493** (2016), 508-536.

[123] C-H. Yi, J-H. Kim, H-H. Yu, J-W. Lee, C-M. Kim, *Fermi resonance in dynamical tunneling in a chaotic billiard*. Phys. Rev. E (3) **92**(2) (2015), 022916, 7 pp.

[124] H. Zhang, *Statistical properties of chaotic billiards*. Thesis (Ph.D.) The University of Alabama at Birmingham. 2005. 151 pp.

[125] A.N. Zemljakov, A.B. Katok, *Topological transitivity of billiards in polygons*. (Russian) Mat. Zametki, **18**(2) (1975), 291-300.

Index